Manning the Future Legions of the United States

Manning the Future Legions of the United States

Finding and Developing Tomorrow's Centurions

Donald Vandergriff

Contemporary Military, Strategic, and Security Issues

PRAEGER SECURITY INTERNATIONAL
Westport, Connecticut • London

Library of Congress Cataloging-in-Publication Data

Vandergriff, Donald E.
Manning the future legions of the united states : finding and developing tomorrow's centurions / Donald Vandergriff.
p. cm. – (Contemporary military, strategic, and security issues, ISSN 1932–295X)
Includes bibliographical references and index.
ISBN 978–0–313–34562–3 (alk. paper)
1. United States. Army—Recruiting, enlistment, etc. 2. United States. Army—Personnel management. 3. Military art and science—United States. I. Title.
UB323.V36 2008
355.2′230973—dc22 2008029441

British Library Cataloguing in Publication Data is available.

Library of Congress Catalog Card Number: 2008029441
ISBN: 978–0–313–34562–3
ISSN: 1932–295X

First published in 2008

Praeger Security International, 88 Post Road West, Westport, CT 06881
An imprint of Greenwood Publishing Group, Inc.
www.praeger.com

Printed in the United States of America

The paper used in this book complies with the Permanent Paper Standard issued by the National Information Standards Organization (Z39.48–1984).

10 9 8 7 6 5 4 3 2 1

The U.S. soldiers, who carry the burden of freedom on their backs

Contents

Preface

"People, ideas and hardware, in that order!"[1]

Colonel John Boyd

Like the United States today, Rome faced multiple challenges in 107 B.C., and was hard pressed to field adequate forces; the number of men who were qualified to serve and who could equip themselves was running out. The Jurgurthine War in North Africa had been going on far too long for the liking of the Roman Senate, a task that counsul (general) Gaius Marius took upon himself to resolve. German tribes had already defeated several Roman armies and threatened Gaul (southern France) as well as Italy.

Marius was a man of vision and acted upon the need to secure Roman provinces with the resources at hand. He did not have a technological revolution at his disposal to solve his strategic problem. Marius turned to an intangible solution: the way the Roman Army manned, structured, and fought its legions. The first thing he did was address how the legions were manned (later referred to in this book as recruited, trained, and retained), and he admitted men of the lower classes. They were recruited to serve long-term obligations of as much as 20 years or more. Then, they were trained, armed, fed, housed, paid, and offered the opportunity of spoils of war. More importantly, for reasons of retention, those who survived the long years of arduous service were given a pension. Given the alternatives these men faced, this was the best they could expect from life. Yet, an even larger personnel change was the ability to promote men from the ranks, who through performance in combat, earned leadership positions, leading others throughout the legion. Through these unprecedented actions, Marius gave Rome what was needed most, a professional army that would expand its borders and provide internal security for centuries to come.[2]

Marius then addressed the force structure (how one is organized) of the Roman Army. The existing system, in which a legion of 4,200 men, most of whom supplied their own weapons and armor and were formed into maniples of 120, was completely overhauled. The new century contained 80 men, and was commanded by a centurion; six centuries formed a cohort, 480 men. The new legion was made up of 6,000 men—nine cohorts of the standard size and one of 800, as well as mounted messengers, cooks, doctors, siege engineers, and other noncombatants. The exact numbers varied over time, but this basic composition was Marius' notion of the structure of a professional army.[3]

Continuing the focus on nontechnological solutions, Marius focused on cohesion and pride. To promote duty and honor, every cohort had its own military standard, and each legion carried a silver eagle, symbolizing Rome. "The soldiers gave devotions to their Eagle standard which symbolized their collective spirit . . . it not only worked to increase the loyalty and devotion of soldiers to the unit and commander; but, it is also reflective of the merging of the old class divisions within the army facilitated by the increased use of the cohort." Slow and cumbersome baggage trains were abandoned; each soldier carried his own equipment, which gave rise to the expression, "Marius' Mules."

The one technological change of note was an alteration to the pilum (spear), whereby a wooden rivet was fitted, which caused the spear to break on impact, so it no longer could be thrown back at the attacking Romans.

These unprecedented changes did not occur overnight, and in the interim, Rome suffered a devastating defeat at Arauiso, in 105 B.C. There, the Germans faced an army led by two incompetent consuls and inflicted upon them the worst military disaster Rome had experienced in a century. But then, the Germans turned to Spain, rather than strike the Italian homeland, and this strategic decision gave Marius the time needed to prepare his new army.

Marius struck in 102 B.C., first crushing the Teutones at Aquae Sextiae, employing an unexpected attack upon the German's rear. Before the campaign was over, 90,000 Teutones were killed and 20,000 captured, including their king, Teutobad. This was followed up the next year with a confrontation against the Cimbri in Northern Italy. The casualties suffered there ended the German threat for the time being—140,000 slain and 60,000 captured, including women and children. Marius' new army had proven itself convincingly; the pattern was set for centuries to come.

Gaius Marius changed the structure of the legion from top to bottom, defying tradition, and created the juggernaut that would build a vast empire that paid tribute to Rome. He changed the face of warfare with this introduction of a professional army through *parallel evolution*, or the changes to several institutions that compose an army simultaneously over time in reaction to evolving enemies. If Marius' reforms had not taken place, the Romans might have been viewed by history as just another European people who struggled to survive against the threats of their rivals. I consider this the first true Revolution in Military Affairs (RMA), or what I call a Revolution in Human Affairs.

Like the reforms of the Roman Legion by Gaius Marius, this book looks beyond what we term recruiting. It is a holistic view of today's Army and makes the case that in order to effectively recruit the soldiers and leaders of the future, the nation needs to take the Army—its personnel management system and structure (how it is organized)—from the Industrial Age into the Information Age.

The Army must be prepared to fight and win Third and Fourth Generation wars. Simply recruiting the right people and then having them step into an antiquated organization means that many of these people will not stay (called retention) as they come into conflict with the premise of responsibility with authority, but instead their ability to contribute and develop is limited by the nature of a top-down, centralized Industrial Age and hierarchal organization. Today, recruiting and retention data bear this out.

In terms of demographics, pure numbers, the Army should have no problem in recruiting and manning its future legions. The same goes for finding and retaining its officers.

Then what is the problem?

There are three answers to the question, and in order to address this complex issue of manning our future Army, one cannot merely address recruiting in itself. As I learned from the Jesuits while teaching at Georgetown University, strands of knowledge are interrelated. Simply put, you cannot address one without addressing other issues. What is occurring with recruiting today is that an Industrial Age organization—the Army—is throwing the age-old solution at the problem. That is, more money. This, of course, will not work, as demonstrated by the backlash of many officers when the Army proposed offering $20,000 bonuses to them if they would stay in the Army.[4]

The first answer to the problem of recruiting for our future legions is that we are now in the fifth year of an unpopular war waged by a society that its leaders have divorced from the war. The other two problems deal with culture and structure. Throwing money and lowering standards in order to man our nation's legions will not solve the problem. They are a temporary fix that the great management theorist Frederick Taylor would be pleased with—because it meets the numbers needed—but it has dire strategic implications. When these legions are committed to today's complex environment of combat, tactical blunders will become strategic liabilities.

In a recent article, "The Army We Have," in the *Atlantic Monthly*, author Brian Mockenhaupt concisely describes the first issue, U.S. society:

> Since the end of the draft, in 1973, the U.S. population has grown by almost 100 million. Meanwhile, with the end of the Cold War, the active Army shrank from 780,000 members in 1989 to fewer than 500,000 in 1996. (The Army expects to have 512,000 soldiers at the end of 2007; Gates's plan would raise that number to 547,000 by 2012.) And many more jobs within all the services have been opened to women, who now make up about 15 percent of the Army.[5]

As I describe in Chapter 3, trends in U.S. society are stacking the deck against recruiters, who are some of the hardest working soldiers in the Army. Mockenhaupt, citing U.S. Army data, goes on:

> In the prime age group for recruitment (17 to 24 years old), 7 in 10 are ineligible for military service, Army officials say. More than half the members of this youth cohort are disqualified for moral, mental, or medical reasons:

> They have had too many run-ins with the law, or they have gang-related or extremist tattoos; they have had psychiatric treatment for severe mental problems or antisocial behavior; or they have been diagnosed with one or more of a staggering list of medical conditions, from heart murmurs to obesity. Other potential recruits have too many dependents, scored too low on the Army aptitude test, or lack high-school or general-equivalency diplomas. Take out those already serving or joining other branches, those who are disclosed homosexuals, and those who are smart and healthy but have no intention of ever entering the military, and the pool shrinks further. From 1976 to 2001, the number of male high-school seniors who say they will definitely join the military remained constant, at about 10 percent. But those saying they would definitely not serve has risen, from 40 percent to 60 percent.[6]

The late Colonel John Boyd (USAF ret.), America's most creative military thinker and theorist, was a self-taught mathematician and aeronautical engineer. His energy maneuverability theories revolutionized the design of fighter aircraft since the 1960s. Boyd used to drill a "machines don't fight wars, people do, and they use their minds" message into the heads of his protégés ad nauseam. Boyd's message, which is central to understanding the Army's fascination with the Theory of Management Science applied to warfare, which has permeated how the Army manned its legions—how it recruits, trains, educates, and manages its people—was that technology is subordinate to and serves people. To understand what technologies work and do not work in war, one first has to understand how people think and act when the fog, fear, and chaos of combat inhibit vigorous activity.

More importantly, it is essential to understand that the very way the Army recruits, trains, and retains its people—both enlisted and officer—is outdated. To the credit of the Army, it is now acknowledging this and appears to beginning movement to fix it; but it is going to take time to evolve its culture. The current culture—its personnel system, which shapes the culture, is based on an organizational model adopted between 1899 and 1904, and an officer personnel system that evolved from the need for rapid mobilization in order to fight the Soviets on the plains of Europe—needs to change dramatically to be prepared to confront the challenges of the future.[7]

Today, the Army's cultural mind is beginning to grasp the meaning of Boyd's commonsense wisdom. Technology was always seen as an end, not a means to an end. The newest counterinsurgency manual, *Field Manual 3-24 Counterinsurgency* (FM 3-24, December 2006), is challenging the emphasis on technology and placing the importance of the human dimension (the mind) at the forefront. Past doctrinal manuals and concept briefs said technology—especially highly complex, expensive technology—would revolutionize the conduct of war until conditions in Iraq and Afghanistan challenged and defeated this belief.

There is a lot of work to be done, but several leaders throughout the Army are beginning to change this cultural attitude. It is going to be hard as well because the historical reliance on technology as a substitute for a truly professional army is ingrained in a culture that prides itself in leading the rest of the nation with its businesslike practices, especially in the way it manages its personnel system. The subordination of soldiers to machines has evolved over the past 104 years into individual parts of an ever-larger bureaucratic machine. The belief that machines fight wars and people are of secondary importance was exemplified by technological solutions to the ongoing enemy evolutions in the war in Iraq and Afghanistan. The high priests of technology in the Pentagon and industry (and their wholly owned subsidiaries in the media and think tanks) even have the temerity to construct a precisely defined vision of a high tech world in 2025 through the RMA. It justified the past obsession with "revolutionary" precision-guided weapons and all-seeing, all-knowing command and control systems. Today, these same systems have not been able to defeat a third world insurgency. Only thinking leaders and soldiers can do that.[8]

Notes

1. John Boyd, "A Discourse on Winning and Losing," unpublished briefing (Washington, DC: August 1987), pp. 5–7.
2. Retrieved from http://www.roman-empire.net/army/army.html, accessed June 21, 2007.
3. Adrian Goldsworthy, *The Complete Roman Army* (London: Thames & Hudson, 2003), pp. 42–60.
4. Colonel J. B. Burton, "Command's Feedback on Officer Critical Skills Retention Bonus," downloaded from the Internet (Iraq, 1st Cavalry Division, May 24, 2007).
5. Brian Mockenhaupt, "The Army We Have," *Atlantic Monthly* (Washington, DC: Atlantic Monthly Publishing Group, May 2007), p. 5.
6. Ibid., p. 5.
7. Samual P. Hays, "Introduction," in *Building the Organizational Society*, Jerry Israel, ed. (New York: The Free Press, 1971), p. 3.
8. Franklin C. Spinney, "Pork Barrels and Budgeteers: What Went Wrong with the Quadrennial Defense Reviews?" in *Strategic Review* (Boston, MA: United States Strategic Institute, Fall 1997), pp. 29–39.

Acknowledgments

A lot of people assisted me with this study, taking the time to contribute their thoughts or to edit, or working with me, allowing me to put my ideas to practice, or just setting an example for me to emulate. For this I am grateful. Of particular note, I am indebted to BG Sean Byrne, Mr. Robert Byrce, MSG Mike Chaisson (USA, ret.), Major Christopher M. Coglianese, Dr. Arthur Coumbe (Colonel USA, ret.), Major Andy Dziengeleski, SGM Ernest Edwards, MSG Rob Frye (USA, ret), Major Robert Goodfellow (USA, ret.), Major Marty Klein, Major Robert Krumm, MSG Bill Lewis (USA, ret.), Mr. Mark Lewis, Mr. William S. Lind, Colonel James McGaughey, Dr. Charlie Moskos, Major Shane Mullins, Colonel Christopher Paparone, Major Darryl Perry, Colonel George Reed, Dr. Chet Richards (Colonel USAFR, ret.), Dr. Jonathan Shay, Mr. Franklin "Chuck" Spinney, General Donn Starry (USA, ret.), Mr. John Tillson, LTC Greg Wilcox (USA, ret.), Colonel G. I. Wilson, LTC Ike Wilson, and the 123 cadets and cadre who contributed their thoughts and comments over the past 6 years, particularly Captains Joe Bernard, Pat Fagan, and Anthony Heisler, who kept me constantly updated with their realistic and intelligent insights into today's Army culture. Pat Fagan also gave me incredible and intelligent interviews regarding his experience in Iraq and how this experience was impacted by his development.

Special thanks go to members of the Asymmetric Warfare Group (AWG). One of the best events that occurred in my ongoing endeavor to make the Army better was teaming up with these individuals in the summer of 2007. First, I appreciate Colonel Robert Shaw, commander AWG; LTC Mike Richardson; and CSM Mike Cortes and CSM (ret.) Hugh Roberts for linking me up with several members and projects in order to provide influence on their work through the Adaptive Leader Methodology (ALM) mentioned in Chapter 6. Two other individuals stand out as well: one is Major Kevin McEnery (USA, ret.), who wrote an unpublished paper entitled "Changing Army Culture," which gave great background regarding the influence of General DePuy on today's Army training and culture. And, finally, there is CSM William M. Darwin (USA, ret.). I consider Morgan Darwin one of the best experts on Army training and leader development, how Army leaders got to where they are today, and, more importantly, how to fix them because he has such a thorough understanding of the "why" and American soldiers. Morgan has been a great coach and mentor to me over the past year.

I also owe gratitude to COL Casey Haskins, former commander of the 198th Infantry Brigade at Fort Benning, GA (2006–2008). COL Haskins has been at the forefront of leading a revolution in the way the Army develops its new soldiers. He also provided me with detailed knowledge on how he led his brigade in transforming training. The Army is very lucky that COL Haskins is moving to be the new director of the United States Military Academy's Department of Military Instruction, which influences all cadets at West Point.

A few other people deserve special thanks: LTC Allen Gill (USA, ret.), my last boss on active duty, provided an atmosphere that encouraged Georgetown ROTC to become a "learning environment." LTC Gill has also become a good friend off whom I can bounce my ideas and get back great and honest feedback; and he has also encouraged good ideas. Major Paul Wilcox commanded one of the first companies at the Basic Officer Leader Course 2006–2007, and had the moral courage to implement many of the ideas outlined in Chapter 6 under the ALM. There is also LTG Franklin Hagenbeck, former Army G1 (now Superintendent of the U.S. Military Academy at West Point), who approved my stay at Georgetown through my last year in the Army. This allowed me to begin and finish the study, "Raising the Bar," that served as a foundation for this book. Thanks also to General Kevin Byrnes, Commander TRADOC, for his limited time to read an earlier version (seventh edition) of "Raising the Bar," which dealt with changing the culture of the Army, and providing his comments, as well as support.

And while not contributing directly to the writing of this book, I owe a mountain of gratitude to my current Army boss and Director of Army Capabilities Integration Center (ARCIC) Forward, Colonel Rickey Smith (USA, ret.). I also appreciate the support from my contracting supervisor LTC James Bierwirth (USA, ret.). They both had strategic insights and trust in pushing my ideas of leader development throughout the Army. I would not have made the gains I have over the past 3 years without their strategic vision and belief in the ALM model, which originated from the study "Raising the Bar" I did in 2004–05 while at Georgetown University.

My thanks to Mr. Adam Kane, Senior Editor of Praeger Security International, and Ms. Alicia Merritt, Acquisitions Editor, Reference, Praeger, who encouraged me to go through with this project, and remained highly positive while I worked on it.

Finally, there is my wife Lorraine, who sacrificed so much time with me as I worked on first the study and then this book. In order to avoid conflict with my day job contracting with the Army, I had to spend many hours writing, editing, and researching during the evenings, over weekends, and on holidays.

By the way, just in case you're going to ask, "Why is Vandergriff doing this?" as a professional, I see a problem and I am going to recommend the best solution. But mainly it is for our soldiers and their leaders. Soldiers are the ones that have to bear the burden of strategic decisions. They are at the point of the spear in a world that is not getting better. They and the citizens of the United States deserve the best leaders that we as an institution can provide. It is our professional obligation to give them the best, even if they sometimes don't understand or maybe agree with how it will be done.

INTRODUCTION

Tomorrow's World Today

"Machines don't fight wars, people do... and they use their minds."
Col. John R. Boyd, USAF (Ret.)[1]

Two profound environmental changes occurred in the waning years of the twentieth century. First, the Cold War ended suddenly. Second, the dominant features of contemporary conflict began to mutate as a variety of irregular forces around the world learned how to attack the political will of their adversaries while bypassing the traditional strengths of conventional military forces. The threads of this mutation reach far back in time.[2]

The emerging pattern of combat in the last half of the twentieth century (e.g., Vietnam, the Intifada, Lebanon, Somalia, Chechnya, Yugoslavia, Afghanistan, and in Iraq, where the rise of organized non-state entities has been dramatic) is that irregular warfare is now spreading rapidly around the world. Its practitioners learn to amplify their successes and discard their mistakes.

The primary exceptions to this trend—the Arab-Israeli wars of 1956, 1967, and 1973; the Falklands War; the Persian Gulf War, the U.S. defeat of the Taliban and al Qaeda in Afghanistan in October–December 2001 and the drive to Baghdad in the first phase of Operation Iraq Freedom, March–April 2003—are beginning to look more and more like transitional anomalies that usually accompany the end of an era. This new generation of warfare has come to be known as Fourth Generation Warfare or 4GW.

But don't view the generations of war as some type of new war. They are not. The generations of war, four of them, overlap, and should be viewed as a different way to approach the conduct of war in terms of the scope and the size of the physical dimension of the battlefield, as well as in terms of time; and finally, the impact smaller groups of individuals can have on strategic outcomes. As the prominent strategist Dr. Colin Gray says, "War is war and strategy is strategy despite the modifiers."[3]

Another factor impeded the military's response to change. This took the form of a gradual buildup of internal rigidities during the 40 years of Cold War. By 2008, a tightly woven web of mutually supporting rigidities—built upon traditions and myths—make it difficult for the Army to reorient its manning, training, leader development, and recruiting practices to the changes implied by end of the Cold War and the rise of 4GW. Currently, we encounter a constant display of great-sounding buzzwords in speeches and articles, and on PowerPoint slides, alongside some great individual efforts with Training and Doctrine Command (TRADOC) that are beginning to disrupt the current cultures with changes to enlisted and officer training. But there is still a great amount of institutional resistance as the Army tries to transform its culture from 2GW to 3GW, so it can cope with 4GW.[4]

By leaps and bounds, command climates within the Army are evolving. But climate changes are short-lived no matter how positive if the overall culture remains the same and simply tolerates the changes due to the demands of the moment. The U.S. Army is rapidly learning counterinsurgency in Iraq. Yet, most of the learning has been from the bottom up, while the Department of Defense (DoD) and the Department of the Army believe they can mandate innovation. They have slogans in search of ideas, and believe that an organization can have innovation only when it has a problem. The United States now has a problem—4GW. Senior officers must think about concepts for 4GW. They see solving the problem as having worked on tactics, techniques, and procedures (TTPs). The Army needs to develop an underlying theory, to write the doctrine, and then develop TTPs alongside changing the Army's culture. Too often, the Army takes technology, designs an organization to use it, and then develops the concepts; with people considered last. But there is now a bright beginning with Army TRADOC actually sponsoring a Human Dimension study that should help spur change. This study addresses institutions across the spectrum of the Army, anything touched by the Human Dimension.[5]

A changing external environment left the U.S. military-industrial-congressional complex (MICC) high and dry, like a beached, bloated whale through the 1990s, and up to 2008 (as demonstrated by the effort to get "up armored" High Mobility Multipurpose Wheeled Vehicle [HMMWV] or Humvees to Iraq), flapping futilely to pump life into its internal organs while engaging in fanciful struggles to return to an environment that no longer existed. By 2008, the struggle had come to a head in the middle of a presidential election, with cries for more budget increases emanating from service chiefs, retired chiefs, news media pundits, defense lobbyists, and a host of voices in Congress.[6]

The aim of this introduction is to present a variety of perspectives to aid in understanding the deeper causes of the Army's dilemma: It is making an attempt at the greatest changes it has ever tried, in the middle of fighting a war, while further demonstrating the need for the recommendations to evolve the Army's culture. This takes place alongside former Army Chief of Staff General Peter Schoomaker's other 17 "areas of focus" and the new efforts of Army Chief of Staff

General George W. Casey, with changes made to recruiting, training, manning, and leader development as the start to building a new foundation for the evolved culture.[7]

The End of the Cold War and the Question of "Ready for What?"

Contemporary force structures (hardware and organizations), as well as operational doctrines (ideas, style of war, and traditions) are largely legacies of events over the entire twentieth century, although one can discern influences reaching back to the Civil War and the Napoleonic era. The assumptions underpinning the personnel system and how the Army recruits and develops its enlisted soldiers and accesses commissioned officers, on the other hand, extend back to the late eighteenth century, beginning with the widespread fear of a standing army held by the framers of the Constitution.

The changed conditions caused by the end of the Cold War, together with the rise of 4GW, have turned these comfortable historical stabilities on their collective heads. They force our leaders to call for the most fundamental—and uncomfortable—re-examination of the military question, "Ready for What?" in more than a hundred years.

It cannot be emphasized too strongly that the strategic question, "Ready for What?" poses an uncertainty new to our experience. The military strategies in World Wars I and II and the Cold War reflected a general agreement that the central strategic problem facing the United States between 1914 and 1990 was the need to counter power imbalances in Europe caused by the rise, first, of Germany and then the Soviet Union. In this sense, the role of the United States has evolved insensibly into one somewhat like that of Great Britain in the nineteenth century.

To be sure, the centrality of the balance of power question to the United States may not have been so clear to a contemporaneous observer at different points in the twentieth century. During the period between the world wars, for example, that observer faced a distracting isolationist debate in the aftermath of the Senate's rejection of the Treaty of Versailles in March 1920. But while that debate was accompanied by domestic political reactions (e.g., budget reductions and demobilization), it did not affect the strategic outlook of military professionals in America's war colleges, who were trying to imagine and plan for the next war.[8]

A contemporaneous observer also would have witnessed the rise of Japanese power in East Asia during the first half of the twentieth century and the global nature of the communist threat in the second half (particularly in Vietnam and Korea). Nevertheless, looking back at the cacophony of events in the twentieth century, it is safe to say that the isolationist debate, the problem of Japanese power, and the global expansion of communism took a back seat when they came in conflict with the strategic realities posed by the heavy conventional capabilities of German and Soviet armies in Europe, and later the Soviet Union's theater and strategic nuclear capabilities.[9]

The cumulative effects of this evolution caused the United States to abandon completely its traditional nineteenth century organizational premises: no entangling alliances, mobilization instead of a large standing army, coastal and frontier defense, and small-scale interventionist operations oriented toward protecting U.S. interests in the Western Hemisphere, as well as American seaborne commerce.

Between 1950 and 1990, these traditional notions were replaced completely by the dynamics of the bipolar Cold War world: a worldwide network of entangling alliances centering on the North Atlantic Treaty Organization, the maintenance of large standing forces, a massive permanent forward deployment of heavy forces in Europe and elsewhere, and the ability to execute large-scale interventions to counter the worldwide communist revolutionary movement.[10]

These organizational dynamics led to a deeply entrenched foreign policy and military strategy fixated on a stable, bipolar worldview. This insensibly suppressed the physical and psychological effects of regional rivalries and caused each side to establish alliances that might otherwise be considered questionable (e.g., U.S. alliances with right-wing dictators in Central America and Islamic fundamentalists in Afghanistan, because they were anticommunist).

The sudden collapse of Soviet power in 1991and Russia's subsequent, albeit imperfect, struggle to adopt Western political values shattered the simple bipolar orientation that had conditioned American strategy and thinking for two generations. More importantly, it restored the balance of power in Europe for the first time in almost a hundred years. This obviated the requirement for a massive forward deployment of heavy U.S. air and land forces on the European continent.[11]

The collapse of Soviet power also neutralized the easily understood bipolar dynamics of the Cold War. This unleashed a welter of hitherto suppressed nationalist, ethnic, religious, and criminal conflicts, which exploded in a rapidly changing, multi-polar, multicultural political context. Each conflict, being rooted in the unique histories and cultures of the belligerents, defied the generalized logic of the Cold War.

Moreover, these conflicts take place in environments that embody poorly understood global forces, which are pushing many parts of the world into chaos. Among these forces are the pressures of overpopulation, rapid urbanization, accelerating environmental degradation, stark poverty and the widening gap between rich and poor, growing scarcities (including water), the effects of near-instantaneous global communications, and the growing moral conflict between Western materialism and Islamic spiritual values.

What do these changes imply for the orientation of the Army, particularly the way it recruits and trains its soldiers and also accesses (chooses) and develops its leaders?

The Evolving Nature of Warfare

The four generations of war began after the nation-state had become the predominant form of government in Europe and European armies were equipped with firearms. However, firearms during this period (running generally from the late seventeenth through the early to mid-nineteenth centuries) were quite primitive. They suffered from short effective ranges, slow rates of fire, unreliable ignition, and an inability to fire effective exploding projectiles. This, in turn, left room for weapons such as the sword and bayonet.

The extensive use of close-order formations was also required, despite their vulnerability. Close-order formations massed enough firearms on a given frontage to repel shock action by cavalry and bayonet-wielding infantry. They could also maneuver, albeit slowly, and battlefields were still small enough for generals to exert personal control over their men. Decisions were usually focused on either the brilliance or incompetence of one individual, such as Napoleon. Leader development came about by default—a leader learned his trade, was wounded or relieved, or died. Meanwhile, many soldiers suffered and paid the price with their lives.

2GW resulted from a combination of the Industrial Revolution and the fact that firearms had become technically mature. The effective ranges of small arms tripled and their rates of fire increased by tenfold. Artillery was able to rain exploding shells upon enemies the gunners could not even see. These new weapons could now be built in great numbers and at low cost. Widely adopted Prussian-style mobilization systems allowed nations to tap much of their adult male populations for military service. Armies of unprecedented size and firepower began to appear. Steamships and railways could move armies and their weapons and supplies wherever they were needed.

Battlefields grew steadily in size until they covered entire continents. The armies became so large that maneuver became difficult. In many parts of the world, armies could easily anchor both flanks on impassible terrain (such as the ocean or mountains) and thus force their opponents into complete passivity or costly frontal attacks. Armies thus began to rely more and more on their firepower to batter holes in their enemies that their infantry (and later, armor) could then exploit. 2GW became a pounding match in which the side with the greatest (or best-managed) resources always won.

Battle resembled an engineering problem more than a military one. As a result of this fact, seen by many as early as the late 1800s, the U.S. officer corps strove to become professional. In order to be accepted by a neglectful public and nervous Congress, the officer corps adapted the emerging cult of management science as its gauge of professionalism. Following closely behind this were education and training systems based on the theories of Frederick Taylor and Max Weber.

Several dramatic changes took place with the tenure of Elihu Root as Secretary of War from 1899 to 1904. While the legacies have a negative impact today on all aspects of the Army, particularly in personnel management, Root's reforms, based on a misinterpretation of reforms advocated by Emory Upton and scientific management used in large corporations, did enough to enable the United States to fight in World War I.[12]

Nevertheless, World War I, which was the classic 2GW struggle, caused great frustration. Even with good planning and management, most armies exchanged huge losses for trivial gains. Tactical control became highly centralized. Large staffs were needed to produce the complex plans required to coordinate artillery fires with advances by infantry. The inevitable "friction" of war often caused plans to go awry, as it was frequently difficult to implement the last-minute changes that mishaps generated. Nevertheless, in 2GW, God generally was "on the side of the big battalions." Since the Central Powers were markedly inferior to their enemies in both population and industry, they finally had to yield after more than four years of slaughter.

For major industrial powers like the United States, Great Britain, and France, which could count on individually or collectively dominating their likely adversaries, 2GW offered a fairly certain, if expensive, guarantee of ultimate victory. Germany, on the other hand, although a major industrial power in its own right, found itself sandwiched into central Europe by powerful enemies. It had to fight hostile coalitions that could draw on much larger populations and industrial bases. Worse, Germany's geographic location made it vulnerable to economic blockade. Its prospects in a world of 2GW were grim indeed.

Therefore, as World War I was raging, Germany relied on an ideas-based and people-centric transformation in an attempt to win the war. It began to move toward a new form of warfare that would leverage its already highly professional officer and noncommissioned (NCO) corps to offset its enemies' numerical and material superiority. This 3GW (or what is known as Maneuver Warfare) began by addressing the new reality that armies had become so large they could no longer be outflanked in any strategic sense.[13]

However, these armies' newfound firepower had forced them to disperse themselves sufficiently that their front lines were riddled with gaps. Small units could easily slip through unnoticed and attack frontline positions from behind. Better still, they could press on and assault the enemy's artillery positions, command posts, and supply dumps. Above all, they could spread chaos and confusion and upset the carefully laid plans of their 2GW enemies.

In order for a 3GW attack to succeed, command and control had to be highly decentralized, as opposed to the highly centralized command system in 2GW. There was no way that a major headquarters could tell a small-unit leader how to slip through an enemy's defensive positions. Once he was in the enemy's rear, the small-unit leader was beyond close tactical control and had to rely on his adaptive skills. This created a major obstacle in developing a new doctrine for the Germans: How were they going to create or find the "adaptive leaders" to execute the new ideas in battle? Never in military history had so much decision making, or the responsibility one leader could be allowed to have, influencing the overall plan of an Army, been put pushed to such a low level in the chain of command.

From the time of the reforms of the German Army after the disaster at Jena in October 1806, the Germans focused on creating and sustaining a professional officer corps. They realized that if they invested greatly in both tangible and intangible resources, in their leader development, and were patient, they could solve many of their potential tactical and operational problems. As they developed their new system and tried to make it work, they realized over generations that their culture had to change and evolve as well; and that this would take generations.[14]

So, beginning in 1809, dramatic and revolutionary policies were implemented. German higher headquarters moved from procedural to directive control. They ensured that they had laid the ground work, supported the culture, so leaders would be educated and trained in order to evolve into being "adaptive leaders."[15]

The Germans realized that not everyone could be a decisive adaptive leader on the battlefield. So although they opened the ability to become an officer to most of the population, they also advocated strenuous selections up-front for those aspiring to or thinking they wanted to be an officer. Even during war, the Germans maintained high standards in leader selection and development—in deciding who had the potential to make decisions, especially under the stress of combat. The Germans began a tradition of "better to have quality than the numbers planners projected they needed to fill their force structure."[16]

The Germans also stepped back and realized that these new adaptive leaders had to have a cognitive ability to understand the overall plan and their role in accomplishing it. After a hundred years, the Germans had evolved their officer corps from one in which entry and rank were based on social status to one in which almost anyone who could meet the tough entrance standards could become an officer, resulting in the most professional officer corps in the world. This evolved later to the German NCO corps as well (many of whom became officers if they qualified upon mobilization for war).[17]

Although the Germans had adopted the doctrine of Maneuver Warfare at the operational level, a culture also evolved beginning in 1809 that allowed for and nurtured bottom-up approaches at solving tactical problems. In the first three years of World War I, the Germans combined ideas with technology to solve the problem of breaking through the trenches, barbed wire, and machine guns, supported by massed indirect fire. Several experiments, by junior officers and their NCOs, had infiltrated to the top levels of the German Army.[18] The German high command then turned around and moved to establish Maneuver Warfare at the tactical level as well. Once they did, the effects of this kind of warfare could be devastating at tactical and operational levels.[19]

Freed from close control and not tied to an artillery fire plan, the attacking infantry, led by adaptive leaders from the *Sturm-strossen* level up to the Army level, literally ran circles around defending forces and created rapidly changing situations to which 2GW forces had a hard time reacting. In 1917, German 3GW offensives launched at Caporetto, against the Italians, and Riga, against the Russians, proved devastating. The Ludendorff offensive against the British and French in 1918 initially also was successful—more so than any previous 2GW offensive. However, it was not enough.

The ideas behind adaptive leaders that the Germans knew they needed to break the backs of 2GW armies were not supported by a complete "transformation." They attained their goal of having no equals in creating and sustaining adaptive leaders at the tactical and operational levels. They provided new weapons, making the squad, with its new leadership, a powerful unit on the battlefield for the first time in history. However, while leadership took off, the rest of the institutions that composed the German Army fell behind and could not support their adaptive leaders.

Horse-drawn artillery and supply services could not keep pace with the infantry's advance, so the offensive ground to a halt before the enemy could be decisively defeated. Despite heavy losses, the Italians, Russians, French, and British all ultimately survived these German attacks. The new ideas had come too late and required too much time for full implementation. More importantly, four years of blockade and bloodshed had worn the Germans down. They were unable to exploit their 3GW successes and ultimately suffered an inevitable collapse.

In World War II, the Germans again employed 3GW, this time enhancing it with tanks, close-air support, and motorization. This proved successful against Poland in 1939, France in 1940, and Greece and the Soviet Union in 1941. A 3GW operation by the Communist Chinese in November 1950 defeated the greatly superior U.S./UN Eighth Army and permanently drove U.S./UN forces out of North Korea. However, despite their successes, 3GW methods alone have not been enough to ensure ultimate victory. The Industrial Age material superiority that dominated 2GW still retained much of its old importance.

Ponderous and inflexible as they usually were, 2GW armies could still defeat their 3GW foes by wearing them down in prolonged battles of attrition. Also, maintaining the fine edge of combat leadership, trust, and cohesion that successful 3GW operations required became increasingly difficult in the face of prolonged combat and mounting casualties. Although it was a considerable improvement over 2GW operations, 3GW retained many features of the 1GW and 2GW styles of conflict. It was based on armies of uniformed soldiers who fought within defined front lines and rear areas and under the control of recognized national governments.

3GW armies still depended heavily on the backing of their national industries and populations, and the industrial superiority of nations like the United States continued to play a decisive role. 3GW methods also were not so easily adapted to naval and air warfare. In those arenas, numbers and technology counted more than anything else. In short, 3GW limited but did not end the military superiority of the major industrial nations. As a result, the victors in World War II saw no reason to update their doctrines to incorporate 3GW methods. They had won without them, albeit at a high price.

Resistance to 3GW methods has remained strong in the United States (except in many parts of Special Forces and in the Marine Corps under General Alfred Gray from 1987 to 1991), despite its defeat in Korea by 3GW Chinese and North Korean troops. Nor had a defeat in Vietnam led to any serious questioning of the "tried and true" 2GW way of war, even in light of the U.S. Army's serious attempts at "parallel evolution" in the 1980s.[20]

However, the emergence of 4GW is likely to have much greater impact. The introduction of nuclear weapons changed everything. Costly as they are, these devices are still far cheaper than conventional 2GW or 3GW armies. Their use, or even the threat of their use, could render the employment of such armies extremely difficult and risky, if not impossible. Conventional warfare has been restricted to regions of the world where neither side would or could bring nuclear weapons into play. As weapons of mass destruction have proliferated, the number, extent, and importance of the regions where conventional armies can safely operate have diminished.

A second important development is the gradual breakdown in the loyalty of individuals and ethnic or other groups to their established national governments. The causes for this are many and varied, but prominent among them is that nation states are losing their reason for being. Modern nations evolved in large part as an efficient means of marshalling the human and materiel resources needed to wage war. However, nuclear weapons are making 2GW and 3GW impossible. Moreover, the welfare state, which attempts to buy popular loyalty with promises of financial security, has diverted so many resources from the military that the gigantic land armies that European states employed prior to 1945 have become largely extinct. This has started to threaten the materiel superiority on which the world's surviving 2GW armies depend.

A third important development, which occurred after the fall of the Soviet Union, is the computer revolution and the creation of the Internet. This has resulted in a change in the nature of wealth and power. In the past, wealth was mainly tied up in land and in capital equipment. Such wealth was not easily moved to avoid taxation, regulation, or confiscation by whatever government happened to control its location. Today, wealth increasingly exists in the form of electronic information. Transmission through the Internet makes this wealth highly mobile. Strong encryption makes it difficult for governments to interfere with the flow of wealth as it crosses their borders.[21]

The Internet also has made it possible to conduct business from almost any location with telephone service. This has left nations like the United States in a vulnerable position. Five percent of American taxpayers earn about 30 percent of America's wealth but pay 51 percent of its taxes. If another country were to offer a haven to these people where their human rights and property would be respected, it could probably attract wealthy taxpayers in great numbers. This would severely undermine the finances of the United States and many European countries whose governments also rely on a small minority of their taxpayers for the bulk of their tax income. National governments would have to compete with each other for taxpayers. In order to do that they would have to become more efficient; they would no more be able to afford expensive militaries than was "cradle-to-grave" socialism.[22]

These developments are starting to divert wealth and power away from nations and toward private and other non-state entities. Nearly all the armed conflicts occurring after the fall of the Soviet Union have involved a nation state on only one side—usually the losing side. Although most nation states enjoy far greater resources than most non-states, the non-states are learning to fight effectively with limited resources. They usually present few, if any, important targets vulnerable to conventional attack, and their followers are usually much more willing to fight and die for their causes. They seldom wear uniforms and may be difficult to distinguish from the general population. They also are far less hampered by convention and more likely to seek new and innovative means to achieve their objectives.[23]

Although the era of 4GW is only beginning, it is already apparent that size and resources, however vital they may have been in the past, can be a liability in 4GW. This is because of the overarching importance of the need to minimize a nation's target profile. All successful 4GW belligerents thus far have practiced this principle. The side that has nothing to attack inevitably forfeits the initiative if it cannot stop its opponents from attacking anything. In Mogadishu, Gen. Muhammad Farrah Aideed successfully denied the U.S. Army's Task Force Ranger a target by constantly moving his headquarters. When Aideed was given an opportunity to attack them, the result was the collapse of the UN intervention.

Fortunately, adjustments were made in the U.S. military special operations command, with amazing results in Afghanistan in October–December 2001 and then in Iraq, particularly northern Iraq in March–April 2003. Yet the revolutions conducted and achieved by Special Operations Command with flattened and matrix organizations, as well as highly professional forces, continued to be burdened by a larger culture, with a force and command structure that continues to trace its roots to the Progressive era and the Napoleonic Age of warfare.

The 1993 street battles in Mogadishu, Somalia, the failure to use conventional forces to enclose the encirclement of al Qaeda forces at Tora Bora mountains of eastern Afghanistan in December 2001, the March 2002 assault on hundreds of al Qaeda fighters and other jihadist guerrillas in Afghanistan's Shahikot valley, and the numerous operations in Iraq and Afghanistan from 2003 to 2008 have too much in common: ad hoc mixes of foreign, conventional, and special operations forces, confusing command-and-control regimes, an underestimated and determined foe, poor battlefield intelligence, and downed helicopters with wounded soldiers whose predicament threatened to foil the entire mission.[24]

4GW thus threatens to place 2GW powers in the position of the elephant terrified by a mouse, of the dinosaur whose eggs are eaten by rodents, or of the lion stung to death by a swarm of wasps.[25]

The Money Is Not Going to Be There?

The Department of Defense's (DoD's) hardening of the arteries—remaining a 2GW power—has its roots in the 40-year semi-mobilization for the Cold War, and in a history of having to raise armies almost overnight to meet crises. While defense budgets and the intensity of Cold War politics waxed and waned between 1950 and 1990, a stable consensus on the existence of the Cold War threat resulted in what amounted to a permanent semi-mobilization. This steady-state condition led to an insensible rise of a different domestic political economy supporting the military than had existed in the United States prior to 1950.

By the time the Cold War ended in 1990, the new political economy took the form of a vast spiderweb of well-developed, intricate relations among defense contractors, politicians, members of the DoD, and a wide array of supporting actors, including publicists, lobbyists, academics, and journalists. Today, this web of influence has gone well beyond the military-industrial complex President Dwight D. Eisenhower warned about in his farewell interests on January 17, 1961, and is more accurately described as a MICC.[26]

Perhaps the greatest threat to U.S. national security is a failure as a nation to strike an intelligent balance between social infrastructure on the one hand and legitimate defense requirements on the other. That there is such an issue at all will surprise many readers, since this is the richest country the world has seen and there is no legitimate foreign threat. Yet, the United States has 42 million citizens without health insurance, and the World Health Organization recently rated the U.S. health care system thirty-seventh in the world (noting that conditions in some of our inner cities and rural areas approach those in sub-Saharan Africa).

In spite of this, service leaders in the Pentagon, politicians in the administration, and certain members of Congress are demanding that the United States spend even more in light of the September 11, 2001, attacks and the ongoing campaigns in Afghanistan and Iraq. The current POMs (service spending proposals), for example, show the budget exploding in 2006, 2007, 2008, and 2009—costs will far exceed the president's fiscal guidance unless (1) the president increases guidance or (2) the DoD "balances" the books with negative wedges (which is more likely).[27]

Budgets will have to increase beyond 2009 if the current force structure and modernization program are maintained. As strange as it may seem, even these plans would not substantially modernize our forces or provide for restoration of lost readiness, since much of the increase goes into procuring small numbers of Cold War–era weapons.[28]

Implications: 2008–2015

The United States may be in the final years of unprecedented prosperity, when hard choices between guns and butter can be postponed or ignored entirely. If the country does go into recession in 2008–09, with unemployment approaching 10%, the numbers of uninsured exceeding 60 million, and the budget sliding into deficit, the president and Congress may come under enormous pressure to address these social issues by reducing defense spending. Needless to say, this would wreak havoc on an already risky procurement plan and cause modernization and readiness to sink to unprecedented levels. Even if the economy stays relatively robust, the United States will, within the next decade, be forced to confront the politically "untouchable" areas of Social Security and Medicare, which will begin to go into the red as soon as 2015.

Although the increasing complexity of weapons and the accompanying growth in cost have been the norm since the mid-1950s, these mutually reinforcing trends accelerated during the 1970s and 1980s with the development and fielding of a new generation of Cold War weapons in the aftermath of Vietnam. The weapons the U.S. military uses today in Afghanistan and Iraq cost far more to procure and operate than the weapons they replaced.[29]

During Vietnam, the defense budget reached a peak of $439 billion (in inflation-adjusted FY 2009 dollars). This budget supported about 550,000 troops in Vietnam, but it also kept hundreds of thousands of other troops forward deployed in Europe, Korea, Japan, the Philippines, Okinawa, and Guam; it funded a rotation in base in the United States to support these forward deployments; and it funded hundreds of nuclear warheads on alert in missile silos, submarines at sea, and airplanes in the air. Now compare this commitment to that of Iraq: To support the wars in Iraq and Afghanistan, the United States will have a larger budget than at the peak Vietnam year, even if one removes the effects of inflation. The FY 2008 budget will reach between $518.3 billion (in comparable FY 2009 dollars), once you factor in the $70 billion supplemental now in preparation, yet our military is only about one-third the size of that fielded during Vietnam. Additionally, if you add $17.1 billion from the Department of Energy, as well as $5.7 billion for the General Services Administration's National Defense Stockpile, the Selective Service and the FBI's international activities, all of which the Office of Management and Budget includes in its so-called "National Defense" budget category, it comes a total of $611.1 billion.[30]

Weapons are projected to age even faster than during the Clinton Administration as the Bush Administration cut back future production (e.g., Joint Strike Fighter). The Army is clearly overstretched by a deployment of only 150,000 troops to Iraq (through the strategy labeled the "surge") and about 15,000 deployed to Afghanistan, as evidenced by the coercive personnel retention policies, such as "stop loss."[31]

Militarily, America's forces are stretched too thin in Iraq, where they are bogged down in a self-protection mode much like the Turks in the Middle East during World War I. And while basic needs of the fighting troops are not being met in a war of choice—armor plate for HMMWV (or Humvee) being the most infamous—DoD continues to merrily throw money at all sorts of Cold War–inspired weapons (e.g., new attack submarines, ballistic missile defense, the F-22 and Joint Strike Fighters, the V-22 Tilt Rotor, etc.) that cannot possibly alleviate the situation for American troops enmeshed in a 4GW.[32]

The continued acceleration in the growth of complexity and cost that created an Iraq War–funding crisis is consistent with two conclusions about the roots of paralysis. First, these internal MICC behavioral dynamics operate independently of changes in the severity of external threats and therefore should be considered autonomous forces misshaping the Army. Second, the MICC is behaving like a neo-Madisonian faction in that it is a subgroup promoting its own welfare (in the form of maintaining Cold War business as usual) at the expense of the general welfare of the soldiers serving in Afghanistan and Iraq.

Each of the three wings of the MICC has its own peculiar, albeit mutually reinforcing, pathology and, together, their workings have created a firmly entrenched, horrendously dysfunctional social system that feeds itself ahead of those of the taxpayers and soldiers it purports to serve. The reasons why troops in combat don't the get body armor they need—while the Pentagon and Congress shovel billions to defense contractors for a missile defense system that won't work and is targeted on threats that don't exist—are just a little more complicated than the pundits on the DC talk circuit, the loudmouths on talk radio or cable TV, or presidential candidates McCain and Obama opine.[33]

The Nature of Change in a Military Culture

The nature of the institution being changed lies at the heart of any question about institutional reform. At its most fundamental level, the conduct of war is a clash of independent wills that operates in the moral, mental, and physical domains. War is not a mechanical phenomenon of physics or technology. War is a living phenomenon, evolving continuously through the interaction of competing human minds with chance and necessity. Moreover, the conduct of war is a group activity embodying multiple interactions among complex formal hierarchies of individuals, each operating according to its own tempo and rhythm under conditions of stress and uncertainty. Organizational culture, therefore, is of paramount importance to the performance of military institutions.[34]

A common culture is the harmonizing glue, the shared set of beliefs, values, traditions, and experiences. It makes it possible for the complex hierarchies to take the initiative rapidly or react with a variety of responses to sudden external changes. They will not fly apart in a struggle to gain sustenance, avoid danger, and overcome obstacles and enemies.

When addressing cultural change, it is also important to recognize that an Army culture does not exist in isolation but interacts continuously with its supporting domestic environment, as well as with the menacing environment posed by its external threats. Finally, the Army and society must remember that the question of how an Army culture changes is driven by the fact that, like all other forms of life, it is a complex, adaptive, goal-seeking phenomenon.

Only a detailed comprehensive plan in which nothing is sacred and where everyone is allowed to talk about each factor in the context of professionalism will show the Army the way to cultural evolution. *Parallel, systematic evolution* is well established in mathematics, physics, biology, and anthropology as the concept that any interaction embodying the characteristics described earlier evolves unpredictably through time via a process known technically as co-evolution.

Parallel, systematic evolution necessarily embodies positive (amplifying) as well as negative (damping) feedback control effects. In the cultural co-evolution of a military organization, for example, the environment (including its domestic dimensions, such as the political and economic institutions sustaining the military) shapes the Army culture while it feeds on and shapes the environment that sustains it.

An example of what Army strategic leaders have to fight is the self-serving use of inefficient, politically motivated weapons procurement strategies. These are motivated by a desire to spread subcontracts around the country to increase the dependency of congressional districts on the continued flow of dollars, jobs, and profits. Leaders of character will begin to question these practices as they orchestrate scarce resources and manpower to perform an array of missions, and every dollar and minute count.

Among other things, this interaction means that changes in the character of a conflict or one's adversaries can also shape as well as be shaped by changes in an Army's institutional culture. This process of co-adaptation—action and counteraction—is straightforward, forceful, and relatively easy to understand during the conflict in Iraq. But in terms of isolation, where the costs of the war are minimized and the pain of suffering is insignificant, the combination of a speculative threat and the shaping effects of domestic environmental factors such as politics and economics make it far more difficult to anticipate and deal with the emergence of new tactics or operational concepts, or to predict how new technologies will actually perform; most of all, they conflict with an evolving culture.

The Army's culture co-evolves with a changing environment over time. During this process, some characteristics carry forward into the future through tradition and learning, whereas others change and mutate according to chance and necessity. By looking to the past, the Army may be able to catch a partial glimpse of the future. However, the interplay of chance with necessity means that the future remains inherently unpredictable. Those who rely on computer-generated global visions of the future to guide current decisions have a different, and an erroneous, concept of change. They fallaciously assume that the Army can plan for the future precisely, so that the interplay of chance and necessity will have little or no impact on developing the future military. As conservative economic philosopher F. A. Hayek has demonstrated, this kind of assumption leads to a fatally flawed argument in predicting how groups of individuals will act.[35]

Institutional change must be dictated from the top down as well as the bottom up. Additionally, strategic leaders must work in perfect harmony with the visionary, while also educating external influences such as Congress. Change does not happen in accord with predictable rules of cause and effect; it evolves from the interplay of chance and necessity. It is a bottom-up process—as is occurring at the hands of junior officers as they cannot wait for their seniors in developing company.com, or the infusion of new soldier and leader development approaches discussed in Chapter 6, "Training (and Educating) Tomorrow's Soldiers and Leaders"—shaped by competition in the free market of ideas as moderated by general guidance.[36]

How to Change

What should the American people expect from Congress and the Army as the United States begins to adapt to this changing face of warfare?

Retired Air Force Colonel John Boyd, a leading military theorist, stated that effective military systems prioritize the components of which they are composed:

people, ideas, and hardware. In the United States, the people aspect is most important. Boyd responded to the Army's emphasis on synchronization—the methodical timing of several events in time and space—with the comment, "you can only synchronize watches not people."

Boyd also emphasized that "people fight wars, not machines, and they use their minds." In sum, military systems that give people top priority adapt to the changing nature of warfare more quickly than those that emphasize machines. Boyd defined this in testimony before Congress in April 1991:

> There are three basic elements [to win wars] and in order of importance they are: People, because wars are fought by people not weapons. Strategy and tactics [ideas] because wars fought without innovational ideas become . . . blood baths winnable or not. Hardware, because weapons that don't work or can't be [produced] in quantity will bring down even the best people and best ideas.

Boyd went on to describe how each aspect is interrelated:

> . . . our military needs to be trained in innovative tactics and strategies that will lead to quick decisive victory at minimum cost to American lives. . . . This requires, first, an understanding of conflict. Conflict can be viewed as repeated cycles of observing-orienting-deciding-acting by both sides (and at all levels). The adversary that can move through these cycles faster gains an inestimable advantage by disrupting his enemy's ability to respond effectively. . . . These create continuous and unpredictable change. Therefore our tactics and strategy need to be based on the idea of adapting to and shaping this change faster than the enemy.[37]

An effective military system is able to combine the concepts that Boyd describes into its military culture.

If this is true, then why has the U.S. Army only slowly evolved itself, particularly now, when so many key people, in and out of the Army, believe it is time to conduct holistic reform? Americans love to boast about their innovation, their ability to adapt and to overcome adversity. If the current establishment is out of date and slow to adapt to the twenty-first century, why can no one, even the Army chief of staff, do anything about it?

This book discusses more than just recruiting. It also addresses the changes needed in the Army culture, by setting the conditions, before introducing changes needed in how the Army creates the "adaptive soldiers" it says it wants—including innovative education and training approaches, as well as how it is structured and manages its people. Effective changes to these institutions enhance the reputation of the Army and not only encourage the recruiting of the right kind of people, but also help retain good people. As a good and brilliant friend, Dr. Steven Stewart, told me recently,

> Your proposed changes to ROTC, how we access our officers, are good and holistic, but they will not go anywhere unless the culture is there for them as they emerge, then it must support-nurture those traits that we know that adaptive leaders and innovators most possess, but many times are in conflict with the reality of today's culture. If an evolved culture is not there to help them, then very little will move forward and change.[38]

I define the Army culture as consisting of more than its actual fighting component. A modern Army, especially one as large as that of the United States, is composed of several interrelated and overlapping institutions. These institutions must evolve together, systematically, to be ready to enter and win in the changing operating environment. I consider the necessary changes to how we create and develop leaders as the most important, alongside evolutions in the culture. To field soldiers and their leaders on the current and future battlefields requires these institutions to work in harmony to ensure that personnel are well educated, trained, and strenuously selected, while employing the right doctrine and their units are equipped for whatever mission confronts them. This is why this book is about more than just recruiting.

Notes

I would like to thank Franklin "Chuck" Spinney, Lieutenant Colonel John Sayen (USMC ret.), COL G. I. Wilson (USMC ret.), Mr. Winslow Wheeler, and Dr. Steven Stewart for their input to this chapter.

1. John Boyd, "Patterns of Conflict," retrieved from http://www.d-n-i.net.

2. Robert Asprey, *War in the Shadows: The Guerrilla in History* (New York: Doubleday, 1975), pp. 16, 28, 222–228. This two-volume study is probably the most comprehensive history of irregular warfare available.

3. Dr. Colin Gray, "*Irregular Enemies and the Essence of Strategy,*" keynote speaker, U.S. Army 16th Annual Strategy Conference (Carlisle, PA: Strategic Studies Institute, April 13, 2005).

4. Colonel T. X. Hammes, *The Sling and the Stone: On War in the 21st Century* (Minneapolis, MN: Zenith Press, September 2004). Readers can find a compendium of articles on 4GW on the Internet at "Defense and the National Interest," http://www.d-n-i.net/second_level/fourth_generation_warfare.htm.

5. Discussion with LtCol John Nagl, "*Irregular Enemies and the Essence of Strategy,*" U.S. Army 16th Annual Conference (Carlisle, PA: Strategic Studies Institute, April 14, 2005).

6. In June 2000, the service chiefs submitted a first draft of the new Future Years Defense Plan that detailed unfounded requirements of $30 billion per year. Several active and retired officers called for an increase in the defense budget to 4 percent of gross domestic product (GDP) in midsummer. If implemented, this budget would exceed the highest levels of the Cold War (see http://www.d-n-i.net/spinney/comments/c386.htm). On September 14, 2000, Daniel L. Crippen, director of the Congressional Budget Office (CBO) told the Senate Budget Committee that the Pentagon's budget needed about $50 billion more per year to fully fund its current program. On September 27, the service chiefs told the

Senate Armed Services Committee that they needed a collective budget increase of about $50 to $60 billion per year to maintain the current force and ensure the future state of readiness (*New York Times*, September 28, 2000).

7. The Army's 17 immediate areas of focus include the soldier; force stabilization; joint expeditionary Army with a campaign-quality capability (formerly called joint and expeditionary mindset); the bench; combat training centers/Battle Command Training Program; leader development and education; current to future force; Army aviation; modularity; the network; Active Component/Reserve Component balance; installations as flagships; actionable intelligence; authorities, responsibilities, and accountability; resource processes; strategic communications; and logistics.

8. Russell Weigley, *The American Way of War: A History of United States Military Strategy and Policy* (Bloomington, IN: University of Indiana Press, 1977). The United States concentrated most of its resources in World War II against Germany, notwithstanding the fact that Japan triggered America's involvement in World War II. Planning for U.S. forces during the Cold War was premised on the assumption that if our forces were configured to cope with the Warsaw Pact/Soviet threat, they could cope with any threats posed by the global expansion of communism. See also John C. Calhoun, *Reports and Public Letters of John C. Calhoun*, vol. 5 (New York: D. Appleton, 1855), p. 84.

9. Donald E. Vandergriff, *Path to Victory: America's Army and the Revolution in Human Affairs* (Novato, CA: Presidio Press, May 2002), pp. 21–56. The Mexican and Spanish-American Wars were anomalies associated with the imperative of imperial expansion (Manifest Destiny and colonialism, respectively). The Civil War, which had a profound impact on American military culture, was an anomaly in that it was an internal affair and largely unrelated to foreign policy nostrums of the founding fathers that guided American policy during most of the nineteenth century.

10. Bruce I. Gudmundsson, *Storm Troop Tactics: Innovation in the German Army 1914–1918* (Westport, CT: Praeger, 1989). Also see Williamson Murray, "Armored Warfare: the British, French, and German Experiences," in *Military Innovation in the Interwar Period*, Williamson Murray and Allan R. Millett, eds. (Cambridge, England: Cambridge University Press, 1996), pp. 4–69. The traditional concept of balance of power is a situation in which no single major European power is more powerful than the combined power of the next two most powerful countries. Williamson Murray, "May 1940: Contingency and fragility of the German RMA," in *The Dynamics of Military Revolution, 1300–2050*, MacGregor Knox and Williamson Murray, eds. (Cambridge, England: Cambridge University Press, 2001), pp. 154–174.

11. Williamson Murray and MacGregor Knox, "The Future behind Us," in *The Dynamics of Military Revolution, 1300–2050*, MacGregor Knox and Williamson Murray, eds. (Cambridge, England: Cambridge University Press, 2001), pp. 175–176. Donald E. Vandergriff, "Why the U.S. Army Cannot Practice Unit Cohesion," unpublished paper presented to the U.S. Army Unit Manning Task Force, November 11, 2002. The author believes the insights shared by General Donn Starry (USA, ret.), during this lecture and subsequent meetings between him and Donald Vandergriff, are pertinent. See also John C. Calhoun, *Reports and Public Letters of John C. Calhoun*, vol. 5 (New York: D. Appleton, 1855), p. 84.

12. Marvin A. Kreidberg and Merton Henry, *History of Military Mobilization in the United States Army, 1775–1945* (Washington, DC: Department of the Army, June 1955), p. 706. Franklin C. Spinney, "Driving Bill and Madeline Bananas," *U.S. Navy Proceedings*, vol. 6, XXIII (April 1999), p. 5. President Dwight D. Eisenhower, "Farewell Address to the

Nation," *Public Papers of the Presidents: Dwight D. Eisenhower, 1960–61* (Washington, DC: Eisenhower Center, 1969); Samuel Haber, *Efficiency and Uplift: Scientific Management in the Progressive Era* (New York: Simon and Schuster, 1964), p. 148. See also Russell F. Weigley, "The Elihu Root Reforms and the Progressive Era," in *Command and Commanders in Modern Warfare*, William Geffen, ed. (Boulder, CO: U.S. Air Force Academy, 1969), p. 15.

13. Weigley, *The American Way of War;* Gudmundsson, *Storm Troop Tactics,* pp. 22–29.

14. Charles Edward White, "The Enlightened Soldier: Scharnhorst and the Militarische Gesellschaft in Berlin, 1801–1805" (PhD diss., Duke University, 1986), pp. 19–26, 33–36.

15. Of course the Germans did not call them this, but instead of relied on the term "Entscheidungsfindung" or leaders of character.

16. In the German army, the Junker cadre of senior officers was extremely clannish, but the lower officer ranks were frequently in close contact with their men. Indeed, the German practice of giving the command of units to relatively junior officers, and even NCOs, made this closeness much more feasible and likely.

17. Martin van Creveld, *Fighting Power: German and U.S. Army Performance, 1939–1945* (Westport, CT: Praeger, 1982), pp. 33–56. Also, Martin Van Creveld, *The Training of Officers: From Military Professionalism to Irrelevance* (New York: Free Press, 1990), p. 134 (U408.3.V36 1990).

18. Gudmundsson, *Storm Troop Tactics,* pp. 43–50, provides the best description of the power, responsibility, and authority, as well as the resources given to engineer Captain Rohr to experiment with new tactics to break the trenches.

19. Ibid., pp. 120–140.

20. See Chapter 4 for a detailed discussion of this term, but in sum it is the evolution of all institutions that compose a military organization in the face of the changing face of war in order to prepare that organization to deal with future threats.

21. Barrie Sherman, *The New Revolution: The Impact of Computers on Society* (New York: John Wiley and Sons, June 1985), pp. 123–131. Joel McNamara, *Secrets of Computer Espionage Tactics and Countermeasures* (Indianapolis, IN: Wiley Publishing Inc., June 2003), pp. 15–34.

22. F. A. Hayek, *The Fatal Conceit: The Errors of Socialism* (Chicago: University of Chicago Press, 1988).

23. Hammes, *The Sling and the Stone.* The decision to deploy peacekeepers to Bosnia is a case in point. The Clinton Administration sold the deployment to Congress by saying it would last 1 year. However, it became clear after the deployment that there was no light at the end of the tunnel. But the American deployment had committed prestige to the operation, which enabled its promoters to then argue that a decision to pull out would destroy U.S. "credibility." This again occurred in Iraq, where the ability to conduct the operation was based on the assumptions that Iraq had weapons of mass destruction (WMD) and that the Coalition forces would be welcomed as liberators, as occurred with U.S. Army forces in France in 1944–45, when they liberated France from Nazi Germany.

24. Sean Naylor, *Not a Good Day to Die: The Untold Story of Operation Anaconda* (Berkeley, CA: Berkeley Hardcover, March 2005). Donald E. Vandergriff, "Why the U.S. Army Cannot Practice Unit Cohesion," unpublished paper presented to the U.S. Army Unit Manning Task Force, November 11, 2002. The insights shared by General Donn Starry (USA, ret.), during this lecture and subsequent meetings between him and Donald Vandergriff are pertinent.

25. LtCol Ralph Peters, "A Grave New World: 10 Lessons from the War in Iraq," *Armed Forces Journal*, vol. 10 (Arlington, VA: July 2006), p. 4.

26. President Dwight D. Eisenhower, "Farewell Address to the Nation," *Public Papers of the Presidents: Dwight D. Eisenhower, 1960–61* (Washington, DC: Eisenhower Center, 1969), p. 103. "This conjunction of an immense military establishment and a large arms industry is new in the American experience. The total influence—economic, political, even spiritual—is felt in every city, every Statehouse, every office of the Federal government. We recognize the imperative need for this development. Yet we must not fail to comprehend its grave implications. Our toil, resources and livelihood are all involved; so is the very structure of our society." "In the councils of government, we must guard against the acquisition of unwarranted influence, whether sought or unsought, by the military-industrial complex. The potential for the disastrous rise of misplaced power exists and will persist."

27. *Ratio of Total DoD Procurement to RDT &E (Research, Development, Testing, and Experimentation)*, presentation (2005). In the late 1980s, DoD began cutting procurement and force structure significantly, while holding relatively constant. The effect has been to aggravate an already severe force aging problem. We may be entering what one Pentagon official recently called a "modernization death spiral," in which the force will consist of a few ultra-modern fighters, and a vast bulge of aircraft 20 or more years old.

28. Franklin C. Spinney, "Defense Power Games" (Washington, DC: Fund for Constitutional Government, 1990). This report can be downloaded from "Defense and National Interest" at http://www.d-n-i.net/spinney/def_power_games_98.htm. "Our military organization today bears little relation to that known ... in peacetime, or indeed by the fighting men of World War II or Korea. Until the latest of our world conflicts, the United States had no armaments industry. American makers of plowshares could, with time and as required, make swords as well. But now we can no longer risk emergency improvisation of national defense; we have been compelled to create a permanent armaments industry of vast proportions. Added to this, three and a half million men and women are directly engaged in the defense establishment. The U.S. is annually spending on military security more than the net income of all United States corporations."

29. Mark Shields, "Our 'Best Equipped' Army? Baloney!" *Washington Post* (op-ed; Saturday, December 18, 2004), p. A27; retrieved from http://www.washingtonpost.com/wp-dyn/articles/A9249-2004Dec17.html. James Madison, Federalist No. 10 (originally titled "The Same Subject Continued: The Union as a Safeguard Against Domestic Faction and Insurrection," *New York Packet*, November 23, 1787). Available on the Internet at http://lcweb2.loc.gov/const/fed/fed_10.html.

30. Winslow Wheeler, "Chew Gum First; Then Walk: Understand Then Contain America's Out of Control Defense Budget" (forthcoming) (Washington, DC: Center for Defense Information, November 2008). Mr. Wheeler also adds in additional costs of military retirement, interest on the national defense, Veterans Affairs benefits, and homeland defense which for national security comes to $863.7 billion.

31. Tom Squitieri, "Army expanding 'stop loss' order to keep soldiers from leaving," *USA TODAY* (Rosslyn, VA: USA Today publishing, January 6, 2004), p. A1.

32. Mark Shields, "Our 'Best Equipped' Army? Baloney!" *Washington Post* (op-ed, Saturday, December 18, 2004), p. A27; retrieved from http://www.washingtonpost.com/wp-dyn/articles/A9249-2004Dec17.html.

33. Franklin C. Spinney, "Defense Spending Time Bomb," *Challenge: The Magazine of Economic Affairs*, July–August 1996, pp. 23–33. This report illustrates the general point

with a case study of air force tactical fighter aviation. It shows how the long-term effects of behavior that drives up costs faster than budgets leads to smaller and older forces and continual pressure to reduce readiness. Finally, it gives the reader an idea of the magnitude of the adjustment now needed to fix the current aging crisis. Promises of lower life-cycle costs have not materialized because the increased reliance on computer diagnostics increased both the variety and quantity of depot-repairable repair parts, requiring a more sophisticated logistics management system to keep track of the growing number of individually accountable items in the supply pipeline. Moreover, by displacing a greater percentage of repairs in space and time from the point of activity, it became more difficult to determine the appropriate mix for war stockpiles. This led to an increasing dependence on wartime work-arounds such as cannibalization to support peacetime operations. All this translates into more money. If, for example, one compares the M1 tank to the M60A3 it replaced, official DoD budget data indicate unit procurement costs increased by a factor of 200 percent and operating costs per mile increased by 70 percent and 180 percent for the M1 and M1A1, respectively. A similar comparison of the F-15 to the F-4 reveals an increase of 240 percent in unit procurement costs and a 53 percent increase in operating and support cost per flying hour. All comparisons have the effects of inflation removed and include the appropriate allocation of depot and replenishment spares costs using official service budget factors. With a few exceptions (e.g., A-10) the overwhelming majority of other weapon categories exhibited a similar pattern of cost increases.

34. Discussions with Major Peter Kline, who developed companycommand.com, February 17, 2005.

35. Discussions with Dr. Steven Stewart, January–April 2005. Dr. Stewart was one of the thinkers behind the Army's School of Advanced Military School (SAMS) and also taught cognitive development at the U.S. Army War College.

36. U.S. Congress, *Congressional Record*, February 23, 1915, 63rd Cong., 3d sess., pp. 4343–4390. See also Andrew J. Bacevich, "Progressivism, Professionalism, and Reform," *Parameters*, vol. 9, no. 1 (March 1975), p. 4. For example, the bomber gap in the early 1950s, the missile gap in 1960, the "window of vulnerability" in the late 1970s, and the inability to prosecute two major theater wars simultaneously in the late 1990s. U.S. Congress gave a $67 million grant to fund productive employment for former Soviet WMD scientists; $30 million of that was spent on salaries of American staffers (including $100,000 golf fees), $20 million was stolen outright by the same American staffers; $10 million was paid to close the program down very, very quietly. The remaining $7 million was spent to employ about 200 Russians for about a year and to report that the number of the Russians employed was 3,370. A Russian producer of bio-weapons (anthrax, small pox) could not get any money to convert itself as the money it requested was spent to buy an apartment for the American program director. Nobody was punished but the whistleblower. Relevant documents are available at http://nunn-lugar.com/def/.

37. As quoted in Major Jeffrey L. Cowan, USAF, "From Air Force Fighter Pilot to Marine Corps Warfighting: Colonel John Boyd, His Theories on War, and Their Unexpected Legacy" (thesis, U.S. Marine Corps Command and Staff College, 2000), p. 14.

38. Discussions with Dr. Steven Stewart, January–April 2005.

CHAPTER 1

Tomorrow's Battlefields?

> "... members of native forces will suddenly become innocent peasant workers when it suits their fancy and convenience."
>
> USMC Small Wars Manual, 1940[1]

War always evolves. Current and potential enemies of the United States are learning and adapting, and the U.S. Army must do the same or lose. But today, war is changing faster and on a larger scale than at any time in the past 350 years. Incremental changes enacted in times of emergency, which then revert back to the old bureaucratic ways in times of peace, no longer will suffice. Not only is the Army facing rapid change in *how* war is fought, it is facing radical changes in *who* fights, what it is fighting *for*, and the type of environment enemies will force the U.S. to *fight in*.[2] To effectively man today's and our future Army, leaders must understand as well as imagine the type of battlefields in which they will be operating.

All over the world, state armies, including that of the United States, find themselves fighting *non*-state opponents in restricted environments. This kind of war, as touched upon in the Introduction and known as Fourth Generation Warfare (4GW), is a very difficult challenge for many complex reasons, but mainly for the war's scope and for the fact that it is a war of wills. More so than ever before, war is one of brains aimed at breaking the moral will of the other side to continue in the conflict.

Is 4GW Just Another Term for "Low Intensity Conflict" (LIC)?

> "We should be the ones in the village, not the people attacking the village."
>
> John Boyd on winning guerilla warfare, *On Winning and Losing*[3]

Simply put, opponents who understand 4GW select strategies that aim at influencing the means—most likely the civil populations—of getting to and influencing the enemy's decision makers. This, of course, presents limitless scenarios depending on the political and military objectives of the numerous sides (more than two can fight 4GW).[4] It is all too easy to misinterpret 4GW. This usually centers on calling it an advanced form of an insurgency or guerilla warfare. The other criticism is the tendency to view generations of war—of which 4GW is the latest—in sequence. Unfortunately, this construct is misleading on several counts.[5]

An article written in 2005 by Dr. Antulio J. Echevarria, entitled "The Problem with Fourth-Generation War," is the latest "attack" or criticism of 4GW. It is the best of the few that exist. His charges against the theory of 4GW can be broken down into four parts: The first is that war does not change in "generations"; his second is that, even if it did, the next generation would not be "4GW."[6] Third, even if 4GW existed, some past wars were also "total wars." Finally, the fourth and most common argument is that 4GW is just another name for insurgency.[7]

Dr. Echevarria's first argument goes

> First, the sequencing of the so-called generations of war is both artificial and indefensible. Portraying changes in warfare in terms of "generations" implies that each one evolved directly from its predecessor, and, as per the natural progression of generations, eventually displaced it. However, the generational model is a poor way to depict changes in warfare. Simple displacement rarely takes place, significant developments often occur in parallel. Firepower, for example, played as much a role in World War II, and the Korean and Vietnam conflicts, as did maneuver; perhaps more.[8]

4GW theorists agree with Echevarria regarding the World Wars as total war. Going back and reading the original article on 4GW, the authors Lind, Wilson, et al., mentioned that some of 1GW survives today.[9] It has been stated several times that the generations of war overlap. Generations are used to portray the impact that a style of warfare had during that time. They also present that style or technique of war that parallels or is a reflection of the type of culture that existed during a certain period. Generations pertain more to time and space—how the given type or technique of warfare impacts the leaders (at all levels), participants, and their supporting societies. That is one reason for 4GW suspicion of a "revolution in military affairs." As Colonel T. X. Hammes writes in *The Sling and the Stone*, warfare evolves messily in stages driven by practical considerations on the ground.[10]

"Firepower, for example, played as much a role in World War II, and the Korean and Vietnam conflicts, as did maneuver, perhaps more." Yes, it is true that the United States dropped more tons of bombs on Southeast Asia than it did in all of World War II. The United States, along with its allies, won World War II, but did not do that well in Southeast Asia. This could suggest that there might have been some new developments in the arena of war.

Likewise, criticism of the "generational" model is unfair. This can be seen in how the "Greatest," "Boomer," "X" and "Y" generations overlap one another. If they were sequential, then everyone in one generation would die before the next generation took its place. Yet under the theory of 4GW, each generation's make-up, as part of an evolving U.S. culture, shares traits similar to others, while at the same time demonstrating differences significant enough to demand a new identity. All the generations of war can employ firepower as a technique, but it is the impact that its use has beyond the battlefield—if applied as the wrong solution to certain problems—as well as the way it is controlled, that distinguishes the different generations of war.

Dr. Echevarria's second argument is

> Second, even if it were valid to portray major changes in the conduct of war as an evolutionary progression from 1GW to 3GW, the next logical step in that progression would not be the sort of super-insurgency that 4GW theorists try to depict. Instead, 4GW would be closer to the vision of Net-centric warfare—small, high-tech forces networked together in a knowledge-based system of systems that enables them to act rapidly and decisively—currently propounded by some theorists. To their credit, the proponents of 4GW criticize Net-centric warfare for being too dependent on high-technology, and for being too inflexible to accommodate a thinking opponent. Yet, and quite ironically, this is the very direction in which the logic of their particular theory of military evolution would lead them, if they were true to it. The logic they use to explain key developments in the conduct of war, thus, actually undermines their case.[11]

To refute the claim just presented, a notable, yet unfortunate, 4GW attack serves as an example of the abilities of a 4GW enemy to attack vulnerabilities of a 2GW military and a larger society that once were considered "safe." A small unit of 19 men (dispersed in four smaller teams) was used to conduct a highly technological attack coordinated using global telecommunication infrastructure. They were also networked—both technologically (see above), ideologically, and socially (the key components and differences from a 2GW-style attack). The enemy acted rapidly. Their attacks were completed before conventional forces (law enforcement) could respond. And finally, they acted decisively—the World Trade Center was destroyed and the Pentagon damaged, and thousands were killed. One might argue about the attack's strategic impact the attack had, but it demonstrates all the traits of 4GW.[12]

The key point is that 1GW, 2GW, and 3GW describe state-vs.-state warfare. Most "4GW theorists" feel that thread has played itself out. 4GW involves state vs. "something else." William S. Lind, for one, defines 4GW in these terms: as a "crisis in the legitimacy of the state." Non-state actors will use techniques such as "terrorism" and guerilla warfare because that is all they have during the early stages.[13]

Third, the charge against 4GW that some past wars were also total wars also misunderstands the scope of 4GW. Opponents claim,

> [4GW proponents] establish a false comparison by which they wish us to conclude that most of the wars of the modern age, which they claim were characterized by firepower or maneuver, were narrowly focused on military power and, unlike the super-insurgencies of the information age, rarely involved the integration of political, economic, and social power. Yet, even a cursory review of the Napoleonic, and the First and Second World Wars, reveals that this is not true. Political, social, and economic capabilities were, in many cases, employed to the maximum extent possible. Some historians, in fact, go so far as to maintain that the First and Second World Wars were, in effect, examples of "total" war precisely because of the extent to which the major combatants mobilized the elements of their national power.[14]

4GW does not mean that there were no "total wars" in the past. Opponents' definition of "total war" excludes 4GW. The major difference between 4GW and past wars is that 4GW seeks to end the enemy's *will* to fight, while previous generations of war focused on removing an opponent's *ability* to fight. The examples of "total war" such as World War I and World War II fall into the latter category. Even they preserved, at least in theory and often in practice, the distinction between combatant and civilian. This distinction seems to be disappearing in 4GW, even as an idea.

Finally, opponents to 4GW usually charge that 4GW is insurgent warfare. They contend that

> In fact, insurgency as a way of waging war actually dates back to classical antiquity, and thus predates the so-called second and third generations (firepower and maneuver) as described by 4GW theorists. Insurgents, guerrillas, and resistance fighters figured large in most of the wars fought during this period. Mao was certainly not the first, nor even the most important, theorist to articulate the virtues of insurgency, or Peoples' war, as it was sometimes called. Clausewitz, for one, called it a "reality (Erscheinung) of the nineteenth century," and provided some valuable insights into its nature. Insurgency did, after all, help the American colonies win independence from the British crown, and it nearly thwarted the ultimate Prusso-German victory over France in the War of 1870–71.
>
> Finally, there is no reason to reinvent the wheel with regard to insurgency as an effective form of war. A great deal of very good work has already been done, especially lately, on that topic, to include the effects that globalization and information technologies have had, are having, and are likely to have, on such movements. We do not need another label, as well as an incoherent supporting logic, to obscure what many have already made clear.[15]

The inclination to use insurgency as a synonym with 4GW is understandable. Guerrilla warfare can be extremely effective. The classical insurgencies were not

4GWs. They were pre-modern wars. Many make the same mistake discussing and using the U.S. wars against the American Indian in the same way. These were struggles based more on family structure than ideological networks; the Indians did not try to influence the decision makers back in Washington. Many of the skirmishes and a few of the battles that occurred during the nineteenth century wars against the Indians of the American West resembled 1GW.

Hardly any of the various Indian tribes were networked, the impact of their few victories only angered the United States more, and the impact of U.S. techniques against the enemy had no bearing on their own or any other nation. The argument that pre-modern war preceded all generations of modern war is meaningless. But put the techniques of 4GW in the hands of intelligent and dedicated transnational groups, and place the focus on defeating state opponents at the moral level (rather than defeating their armies in a Clausewitzian decisive battle), then mix in some modern, off-the-shelf technology and maybe even a stray nuke somewhere down the line, and you might have a new form—dare one say "generation"—of armed conflict.

As was the case in the wars waged by the U.S. Army against the Plains Indian tribes of the American West in the nineteenth century, the armies of states have vast superiority over their non-state opponents in most of what the Army calls "combat-power": technology, weapons, other equipment, techniques, training, and so on. The question, though, is not whether insurgency (more correctly, guerrilla warfare) is new. The issue is what is happening to armed conflict as a result of the proliferation of nuclear weapons, the fall of the Soviet Union, and the emergence of powerful and self-organizing networks that function outside the state system.[16]

In 4GW, despite these superiorities, more often than not, state militaries will end up losing because war now encompasses far more than the ability of one side to out-fight the other in close combat. One reason is because while many of the rank and file try to adapt, the overall organization does not. As Max Boot writes, "If you want to see what lies in store for the armed forces in the future, you could do no worse than to cast your gaze back to the past."[17]

America's greatest military theorist, Air Force Colonel John Boyd, used to say,

> When I was a young officer, I was taught that if you have air superiority, land superiority and sea superiority, you win. Well, in Vietnam we had air superiority, land superiority and sea superiority but we lost. So I realized there is something more to it.[18]

This chapter is about that "something more" that was discussed in a strategic sense in the Introduction. In order to recruit the right people, then prepare them to fight 4GW and win, the Army—its leaders and soldiers, as well as all the U.S. military services—need to understand what that "something more" is. That, in turn, requires an intellectual framework—a construct that helps Army leaders make sense of facts and events, both current and historical. Understanding the "why" enables the Army to create the "what," in terms of what type of leaders, and how to support those leaders, but also in the areas of doctrine, force structure, and technology. However, in order to get to the "end state" of what the Army should look like to deal with 4GW, understanding the operating environment, the "why," also assists in defining the "how to" of a strategic plan to evolve the culture.

The intellectual framework put forward in this chapter is called "Why?"[19] The genesis for these thoughts was first laid out in an article in the Marine Corps' *Gazette* in October 1989.[20] In this framework, modern war begins with the Peace of Westphalia in 1648, which ended the Thirty Years War. Why? The state—which was, itself, relatively new—established a monopoly on war.[21] After 1648, war became something waged by states against other states, using state armies and navies (and later air forces). This occurred first in Europe and then, gradually, worldwide. The Army culture assumes that war is something waged by states and this belief is so automatic that it has had difficulty thinking of war in any other way, until recently, in the last five years of its war in Iraq and in Afghanistan. Indeed, war against non-state opponents is sometimes (misleadingly) called "Operations Other than War" or "Stability and Support Operations."[22]

In fact, before the Peace of Westphalia, many different entities waged wars. Families waged wars, as did clans and tribes. Ethnic groups and races waged war. Religions and cultures waged war. So did business enterprises and gangs. These wars were often *many-sided*, not two-sided, and alliances shifted constantly.[23]

Not only did many different entities wage war, they used many different means. Few possessed anything the Army would recognize as a formal Army, Navy, or Marine Corps. Often, when war came, whoever was fighting would hire mercenaries, both on land and at sea (the Grimaldi family, which today rules Monaco, got its start renting out fleets of war galleys). In other cases, such as tribal war, the "army" was any male old enough, but not too old, to carry a weapon (the Latin word "populus," meaning "people," originally also meant "army"). In addition to campaigns and battles, war was waged by bribery, assassination, treachery, betrayal, even dynastic marriage. The lines between "civilian" and "military" and between crime and war were hazy or non-existent. Many societies knew little internal order or peace; bands of men with weapons, when not hired out for wars, simply took whatever they wanted from anyone too weak to resist them.[24]

Here, the past is prologue. Much of what the Army now faces in 4GW is simply war as it was fought before the rise of the state and the Peace of Westphalia. Once again, clans, tribes, ethnic groups, cultures, religions, and gangs are fighting wars, in more and more parts of the world. Another factor that is not new is the primacy of politics, a protracted nature, and the low utility of strictly military or kinetic answers. This, in turn, drives ambiguity, creating frustration at all levels. But most difficult for the U.S. Army, as well as for U.S. society, is a critical need to understand other cultures.[25] These 4GW enemies also fight using many different means, not just engagements and battles. Once again, conflicts are often many-sided, not just two-sided. Soldiers who find themselves caught up in such conflicts quickly discover they are hard to understand and to prevail in, particularly when operating under the 2GW Army culture.[26]

A "Terrain-Analysis": The World[27]

> "We are preparing for the war we want to fight . . . not for the conflicts we cannot avoid."
>
> "No matter how hard we try to take our world with us, we will still find that we sometimes must fight the enemy on his ground, by his rules."
>
> Ralph Peters, *Fighting for the Future: Will America Triumph?*[28]

Army leaders use a process called "mission-analysis" when planning for any mission. It includes a "terrain-analysis"—a thorough examination of the environment or the "battle-space," where a unit will operate and perform their mission(s). Leaders and their units should be tailored specifically for their mission, the enemy they are going to face, as well as the terrain in which they will operate based on a thorough mission analysis by their high headquarters as well as by themselves. The 4GW "terrain" that is entered by today's leaders and will be entered by tomorrow's adaptive leaders is called "world battle-space."

4GW can encompass a "world battle-space" that includes mismatches in economics, environment, population growth/migration, and distribution of power. These are seen as major *causus-belli*.[29] More importantly, the listed factors are resulting in gaps between the "haves" and "have-nots," and they are growing, particularly when the "have-nots" have access to television and see what they don't have.[30] This is greatly influencing and creating a sovereignty versus anti-sovereignty paradox, which is also growing in modern international politics as politicians are caught up in keeping power by continuing to have to or promising to give more to the populations of the state, which, in turn, takes more away from the "have-nots," bringing the "haves" in more contact with an angry "have-not" population.[31]

Wars are occurring more and more over combinations of environmental, demographic, and political power issues. The poorer people are politically dispossessed and pushed into less valuable land even as their numbers explode.[32] It's only a matter of time before they don't take it any more, and feed the ranks of existing non-state groups, or form their own. Economic trends indicate fewer people "have more" and more people "have less"—and the gap between "more" and "less" is growing, resulting in more wars, especially since this critical issue is all but ignored in the United States.[33]

From a political standpoint, the United States is facing a painful paradox in that it is an institution founded on the uncompromising principles of state sovereignty and freedoms for its people ("human rights"). Yet, it cannot take actions to protect itself or it will violate these very principles, while trying to destroy its enemies in the name of self-determination of its people. States are not supposed to take aggressive actions against states—that is supposed to stabilize the world and bring peace. Unfortunately, the end of the Cold War has brought about more intrastate conflict than interstate war. Trying to grant political autonomy to oppressed people—often to the point of creating their own state—weakens the host states and leads to a proliferation of weak, unstable states. This is not contributing to world stability; indeed, it has precisely the opposite effect.[34]

The U.S. Army is now learning not to respond with armed responses—firepower—where brain power is now required. Yet, the larger U.S. society and the Department of Defense (DoD) readily think that what they have in terms of units and weapons are relevant in today's world. The United States continues to do this without question, though below the surface individuals are beginning to question this Industrial Age approach. A simple analogy is the results of one side adapting while the other remained enmeshed in the tried and old ways during the Hundred Years War. The French thought their knights, armament, and horses were relevant in that environment and did not question it. At Agincourt, French heavy knights got bogged down and were destroyed by mobile, lightly armed foot soldiers armed with the long-bow.[35] Will the United States have to wait for its own Agincourt?

The U.S. Army painfully has learned in Iraq from its similar experience with Vietnam, where the Army thought its heavily armed soldiers, firepower, and mobility were relevant in that environment and did not question it. Doctrine and strategy focused on winning battles, not the war. Then entered North Vietnamese General Giap and 4GW, where war was seen as being fought for decades and the target of the North Vietnamese was the morale of the U.S. public, which, in turn, finally influenced political decision makers.

In regard to technological angles, Moore's law suggests that information technology becomes obsolete overnight. Such a technological race means a perpetual window of vulnerability opens as the U.S. Army struggles to catch up—the Army never gets really good with what its soldiers have before it is obsolete. Yet, this very technology that many claim and seek an answer in has created more vulnerabilities, including a dependency on military contractors. The United States is now dependent on its contractors to support the latest technology. The question continues to arise: Are they combatants or noncombatants? What do they consider themselves? But more important for Army leaders to consider is what do the enemies of the United States consider them? We have already seen the consequences in Iraq. Technological solutions are also dependent on large and long logistics tails, which also increase weaknesses.[36]

The United States is vulnerable to campaigns to manipulate international media and public perceptions. General Aideed proved that he could master U.S. public and presidential perception even without high technology, simply because the United States is so dependent on high technology. Dragging bodies of dead Rangers through the streets of Mogadishu in front of Western television crews did much to end U.S. involvement in Somalia. Yet, in contrast, the televised beheading of captives of insurgency groups in Iraq backfired on those groups—not only failing to damage U.S. public support for the war in Iraq, but turning some of the Iraq population against the insurgents.[37]

Political, economic, and technological climate favors an increase in asymmetric attacks (terrorism) against the U.S. homeland. U.S. vulnerabilities to asymmetric attack lie largely in the civil sector. Indeed, given the immense military strength of the United States, current and potential adversaries have no choice but to see the U.S. weak point because they know they cannot take us on in a symmetrical fight. The enemy wants to find long-bow men to fight U.S. knights in shining, heavy armor. The political, economic, and chaotic technological scene encourages this as the changing face of war. Whether the enemy wants to attack with weapons of mass destruction, a computer virus, or a terrorist act, the best place to hit usually does not lie in the military realm, but in the civilian society and larger civil infrastructure that supports it.[38]

Who is the enemy?[39] Who leads them and who compose their ranks? The United States has the misperception that war is something everyone wants to avoid. However, this feeling is confined to westernized countries that felt the hard hand of total war in the twentieth century. The populations of Western societies are the minority in the world. Most men—especially men from non-Western cultures and less-developed areas—take great pleasure in waging war.[40] A large percentage of mankind actually loves war, loves the violence in it, loves the sport in it. These are people who will not capitulate easily.[41]

How do these non-state warriors fight? How are they organized? These are two questions in the realm of 4GW that are almost impossible to answer due to the varying backgrounds of an array of possible and existing opponents. It is more important to shape the U.S. Army and its leaders with the mindset of adapting to the mission after understanding and defining the threat with questions like "Who is the enemy?" "Where does he come from?" "Why does he 'hate' us?" and "What does he want?" Answers to these questions, along with a picture of the "world battle-space" allow Army leaders to "task-organize" for the mission, in support of the strategy.[42]

Distinction between crime and war is blurring in underdeveloped areas. On top of this, in the underdeveloped and emerging weak states, political power is usually gained through criminal means. If war is an extension of politics and policy by other means, then war and crime are often blurred. Ask anybody what the "war on drugs" is really about.[43]

Today's terrorism is tomorrow's legitimate weapon of war. Most of the U.S. citizens don't know it, but the things they think are terrorist acts today will be accepted as legitimate tools of war tomorrow, and have already been accepted as such by some states and many non-state opponents. Yesterday's terrorist weapons included the submarine and the bomber—both are considered fairly mundane weapons today.[44]

Enemies will likely succeed by waging war between the seams in the U.S. legal system, not against our operational capabilities. The United States is also beginning to experience enemies attacking U.S. laws, values, and perceptions, not U.S. military capabilities. With the rise of information warfare, the United States can expect these attacks to increase. The more the United States focuses on humane weapons—precision guided munitions and non-lethal technologies—the louder U.S. adversaries will advertise to the skeptical world those occasions when they do not work.[45]

To counter this evolution of war, everything the Army does must change or evolve, beginning with the culture, while it retains its core values. The Army was moving to do this as this book was being written, but its biggest adversaries included the larger culture of the DoD and even that of American society. Army senior leaders must convince their superiors for the need for cultural change and garner their and the public's support. The nation must also move beyond rhetoric.

The nature of the Cold War era planning process does not deal with unanticipated radical shifts. As even the former vice chairman of the Joint Chiefs of Staff, Admiral Owen, has pointed out, the military is trapped in an Industrial Age joint strategic planning system that takes a long to do anything.[46] Chairman of the Joint Chiefs of Staff, General Richard B. Myers, said in September 2004 that:

> The U.S. military must overcome the way it has trained and educated leaders. Defense officials are moving ahead to overcome Cold War era training to create a new generation of leaders who aren't constrained by what the doctrine says.[47]

But is it? If this is true, then why is the nation retaining everything from force structures to weapons that have the appearance of a Cold War force?

Soldiers are still being asked to carry the burden of dealing with the changing face of war, while everyone else talks. Everyone knows that it is hard to change the culture, but it is crucial to change the culture to one that is built upon trust, in a flattened, extremely professional military, where the contributions of individuals are valued more than the rank they achieve. The Army is now evolving from training soldiers, and particularly leaders, that the only solution to problems in 4GW is to kill people and break things. The challenge is to retain this ability, but also educate, and then reinforce through training, how to think and find different solutions to complex problems—with killing and breaking things as only one option if the situation calls for it.[48]

The Army Culture and 4GW

Similarly, because practitioners of 4GW will be transnational groups without territorially based armies, much of their activity will probably resemble "guerilla warfare" or "low-intensity conflict." These highly irregular practices have enabled groups that are weak militarily to defeat larger, stronger forces, and they have deep roots in the history of war. The word "guerilla" itself, for example, dates back nearly 200 years to Napoleon's occupation of Spain.[49]

Until recently, however, such "special" operations more often harassed than decided outcomes—they were "sideshows" (as T. E. Lawrence once termed them) in wars fought mainly along 1GW, 2GW, or 3GW lines. Examples could include operations by colonial militias and guerillas during the Revolutionary War, Nathan Bedford Forrest's cavalry raids, partisans during World War II, and the tactics practiced in the early stages of most "national liberation" wars in the twentieth century, including Vietnam. In all of these, conventional forces delivered the culminating blows.[50]

> The move toward fourth generation warfare is occurring in *parallel* with the *move into the information age—i.e., with the political, economic, and social changes affecting society as a whole—and the essential characteristics of this new form of warfare have been clearly illustrated in recent conflicts.*
>
> If we look at the development of warfare in the modern era, we see three distinct generations. . . . *Third generation warfare was conceptually developed by the German offensive in the spring of 1918. . . . Is it not about time for the fourth generation to appear?* (Marine Corps Gazette, October 1989)

There is good news, though, in that an example of how "why" and "what" are evolving parallel with "how-to" in Army basic training and advanced individual training across all military occupational specialties. The evolutions that U.S. Army Training and Doctrine Command are placing upon how the Army trains and prepares soldiers for 4GW, and how they are doing it through the drill sergeants, are a small piece of cultural evolution through *parallel evolution*. This is the end result of the filtering down of the vision moving forward toward the goal. Yet the Army is now evolving the "mental preparation" of its leaders, particularly those who will assume the title, responsibilities, and skills required of the "strategic lieutenant."[51]

With the exception of special operations forces (SOF) and selected command climates, the Army's culture has prepared officers for 2GW, and this legacy still persists. Today's Army includes an out-of-date force structure, tied together with information technology, led by officers (good people all the same) limited in a culture that promotes centralized decisions and stifles subordinate independence and autonomy. The focus has been to fight on the linear battlefield.

2GW advocates the use of massive firepower, calling for a strictly controlled battlefield outlined by detailed graphics. For example, in the first Gulf War, the use of increased firepower in reaction to the enemy was based on our emphasis on teaching checklists and lock-step procedures at our branch schools and combat training centers. A similar force is seen today despite the best efforts by junior leaders and soldiers to innovate to respond to rapidly changing conditions.[52]

4GW is an extension of 3GW with no limits to its depth, no front lines, with targets going beyond the traditional type—that is, military units. 4GW is irregular, with war-fighting skills/capabilities in close-quarters combat and small unit operations among state and non-state actors. In contrast to the U.S. Army's current 2GW-focused doctrine, 4GW calls for a decreased reliance on firepower/attrition in ground combat. It also decreases the reliance on deep strike/strategic bombardment in air warfare. The Army is trying to adjust to this reality, and in many places tactical units have done well, but the entire culture has to nurture this bottom-up change.

Future adversaries, driven by the moral forces of cultural and ethnic differences, are learning how to neutralize the technological advantages of industrial-strength, firepower-intensive armies, particularly in irregular close-quarters combat in urban and suburban areas. In Chechnya, Beirut, Mogadishu, and Baghdad, front lines disappeared; the distinction among friend, foe, and non-combatant became vague to nonexistent, and simple hand-held weapons (RPG-7s), used by well-disciplined, small irregular units, turned armored vehicles and helicopters into coffins and conventional formations into death traps. The first Intifada, armed with stones, reinforced by CNN, bought more for the Palestinians than four conventional wars with Israel. Iraqi insurgents have identified and are attacking the Achilles heel of the United States, its logistics.

While the form of 4GW has roots reaching back at least to T. E. Lawrence and Lettow-Vorbeck in World War I, it is still evolving and is not yet well formed or understood. One common denominator, however, is beyond dispute: The premium on *individual initiative has increased*, enhanced by information technology.[53]

But with this understanding, and a demonstration that the Army can put to execution a small "how-to" of *parallel evolution* through its preparation of new soldiers, the evolution must be applied across the board, and in the way we prepare potential leaders to become officers in a revised accessions process.

Leaders and Soldiers in 4GW[54]

> "The distinction between war and peace will be blurred to the vanishing point. It will be nonlinear, possibly to the point of having no definable battlefields or fronts. The distinction between "civilian" and "military" may disappear."
>
> "The Four Generations of War," William S. Lind, Nightengale, Wilson, et al., *Marine Corps Gazette*, 1989[55]

Today's war validates the need for a U.S. soldier better prepared for the challenges of the future. It will require a "fundamental paradigm shift of how to select, educate, train, and assign Soldiers and their leaders at all levels."[56] One of the Army's leading thinkers in transformational leadership development, Brigadier General (ret.) Huba Wass de Czege, wrote:

> [C]ompany commanders are becoming increasingly important. They ought to have seven or eight years' experience in a battalion before taking command because they need to gain experience to be truly effective. Also, the Army must conduct a more thorough assessment and selection process, choosing only the best and leaving the officers in position longer.[57]

Former Army Captain Mark Lewis has done brilliant work on the need to evolve how the Army prepares its junior leaders for future war. Lewis advocates that if the Army fails to evolve its leader development programs, and the way in which it supports its future junior officers—those from the next generation, "Gen-Y"—then it diminishes the desired effects of its Revolution in Military Affairs. This argument, of course, counters the DoD theorists of net-centric warfare, who see a "system of systems" as the end all of how to fight future wars. As Dr. Fred Kagan has recently pointed out, DoD "has adopted a vision of transformation that relies on high-technology weapons systems rather than on soldiers."[58]

The Army expected to achieve a leap in capability through technology (first) and leader development (second), often using the phrase "Revolution in Military Affairs" or RMA. When diagrammed, there is a slight dip in the "capability" curve early in the timeline to represent transition, but soon thereafter it begins to increase dramatically. That dip is a conceptual representation of the managed risk associated with transformation on three levels. A "threat" line represents the sum of all adversaries and conditions that would challenge the Army, and it is rising—driving the need for transformation.[59]

Back to Kagan's analysis of DoD's net-centric-warfare. Kagan's criticism of the DoD's dominating view is necessary here because it bears down on the Army to follow the same line of reasoning. At least it distracts the Army from its claims of a transformation that is more people-centric. Kagan validates this claim with: it "hinders the conduct of current operations . . . [and] flies in the face of historical lessons about how to transform the military."[60]

Another recent paper on transformation, "To Change an Army: Understanding Defense Transformation," authored by Major Leonard L. Lira, an instructor at West Point, takes Kagan's argument further:

> Kagan points to a defense department authored book, *Network-Centric Warfare* to illustrate how the DOD has been attempting to make the Pentagon more efficient through the leveraging information technology and thereby achieve a defense transformation at the lowest possible cost. Kagan ascribes this strategy to Secretary Rumsfeld's business expertise and background in the McNamara regime. He backs this assertion with evidence that decisions for funding new weapons systems are being based on how they further the current forces' ability to conduct Network-Centric Warfare and systems that bring other capabilities to the force are being less funded.[61]

Lira's thesis actually points to society's lack of understanding of the changing face of war when DoD pursues change in the name of efficiency. This is a theme that has been dominant in the country and its military in finding a technological "silver bullet" since the Elihu Root reforms of 1899–1904. Lira is on target when he

> Concur[s] with Kagan that changing the military with this McNanmarisque economic strategy is not transformational. Seeking to make the military more

> efficient through technology only is a first-order change that will make the military better at one thing, according to Kagan, the ability to "identify, track and destroy enemy targets from thousands of miles away." In the end, this change only reforms the military to meet its current functional imperative given new technology. It does not contribute to changing the core functions, or to multilevel changes I would argue are required to match the new paradigm in which the military needs to operate in; total conflict operations. In fact, this efficiency strategy of the DOD, as described by Kagan does not take the military into a new direction to match the security environment of the 21st century, but rather lays out a path toward fighting the future Soviet replacement of the 21st century.[62]

Lira claims later that Kagan's argument is wrong in that the transformation strategy of DoD is actually a logical reflection of U.S. society's refusal to "fully fund and resource the resources for the new missions and roles that they hold the military accountable for accomplishing." This implies that net-centric warfare is a viable solution to a major restraint placed upon the military by U.S. society.[63]

The common "efficiency argument" is used in choosing the "hardware-ideas-people" course of action so common in DoD. It is more of an excuse for not understanding war as well as the domination of money and major weapons' contractors in transformational decisions. If efficiency was a major factor in decision making, then the people-first argument made so often by John Boyd would not only be seen in decisions, but in results.

While the U.S. military and its society remain focused on technology to solve global military problems, the 4GW environment demands more than ever from today's and future Army leaders. The face of the "battlefield" has changed, creating the conditions for the rise of the "Strategic Corporal" and "Strategic Lieutenant"—a soldier and lieutenant who, fortunately, now are being prepared using less of the out-of-date industrial training model, held in place by out-of-date beliefs, and more via new innovative approaches discussed in Chapter 6.[64]

In the past, Industrial Age (2GW) hierarchal career timelines were assumed to run parallel to the amount of experience (equated to time and rank) needed prior to an officer being allowed to deal with increasingly complex problems. Despite new advances in soldier and leader development, today's career patterns do not reflect reality or necessity, are founded on assumptions made over a hundred years ago in the Industrial Age for 2GW, and will continue to offset and diminish today's advances in training and education reforms.

4GW demands that leaders know more—achieve professionalism—earlier or they fail. The decisions of today's and tomorrow's lieutenants can and will have strategic consequences. Therefore, Army Accessions Command—responsible for the recruiting and preparation of new soldiers and cadets—must prepare its new soldiers and cadets with an extensive base of education, training, and practical experience that can guide them in identifying challenges and developing appropriate responses. The simple application of rote battle drills or standard operating procedures to solve problems will be insufficient, as CPT Pat Fagan says as he reflects on his experiences in Iraq. His preparation began early, while he attended Georgetown University and its ROTC program:

> The tactical training that I received in Georgetown Army ROTC was critical in my success as a scout platoon leader in Iraq from February 2004–July 2004. The tactical decision making process that I learned did more to prepare me for the complexities of 4th Generation Warfare than any other training I have received in the Army. When I refer to the "tactical decision making process," I do not speak of a set checklist I used derive the solution to any problem. Rather, I refer to a method of making tactical decisions under pressure that I learned and refined using a multitude of training tools while at Georgetown.[65]

Potential opponents are well aware of U.S. Army recent actions—successes and failures—and will use unheard of techniques to attack vulnerabilities of the United States. As the nation has witnessed so dramatically since September 11, 2001, and with the evolving insurgency war in Iraq today, enemies using asymmetric capabilities are challenging the United States at all levels of war. But actions have to meet what the Army is advocating in writing and in PowerPoint presentations.[66]

For example, the Army is stating, and some units are practicing, decentralized control and devolution of important decision-making authority to lower levels of the organizational hierarchy, and this has become more evident as units disperse in Iraq according to the new Army doctrine publication *Field Manual 3-24 Counterinsurgency*. The evolution of war has accelerated the need to take theory to reality. Dispersed and numerically smaller formations led by leaders of lower rank will exercise greater influence on the outcomes of land operations. Small units or individual leaders will have the power to dramatically influence operational and even strategic objectives through their actions. Because of decreased mobilization times, these leaders' first battles or other actions may be decisive ones. This leaves less room for learning and adaptation in the midst of such rapidly decided operations.

Two concerns stem from leaders and their units that are increasingly networked, making them more aware. One is that under the present culture, leaders above will micromanage their forces through the enhanced capabilities of new systems (thus frustrating talented junior leaders). The other is that lower level leaders will macromanage a situation since they will have a high degree of situational awareness to use in judging for themselves which actions are appropriate.[67]

Senior leaders will likely choose the former command style if the quality of junior leaders declines, as it did in Vietnam. As a result, an already slow decision cycle burdened by a bureaucratic and bloated officer system will get slower. This is dangerous if junior leaders become isolated with reoccurring breakdowns in advanced technology as it operates longer in combat. Micromanaged and/or poorly prepared junior leaders will be left on their own to make decisions. If they fail to act, or make a bad decision, this can impact all levels of war. The responsibility of our junior leaders is dramatic. Their errors can be reported to the top of the National Command authority and on billions of television sets within minutes.[68]

The affordable solution with long-term benefits to the Army is preparing leaders capable of handling responsibilities that even World War II battalion commanders could not imagine.

Culture vs. Climate; 2GW vs. 4GW

> "The contradiction between the military culture *and* the nature of modern war confronts a traditional military Service with a dilemma. Terrorists resolve the dilemma by eliminating the culture of order. Terrorists do not have uniforms, drill, saluting or, for the most part, ranks. Potentially, they have or could develop a military culture that is consistent with the disorderly nature of modern war. The fact that their broader culture may be non-Western may facilitate this development."
>
> "The Four Generations of War," William S. Lind, Nightengale, Wilson, et al., *Marine Corps Gazette*, 1989[69]

The Army should not confuse *climate* with *culture*. There are great leaders serving in Iraq and Afghanistan establishing the right climate for their subordinates and small units to succeed. The Army is highlighting and trumpeting these successes, as it should—focused on people making positive impacts on the organization. But it should also recognize those units in which command climates, likely influenced by the larger culture, are not evolving and adapting faster than the enemy. Unsuccessful climates cannot be ignored in the hope that positive successes will outlast them, as the negative will have a larger strategic impact than the positive.

Climate is a short-term phenomenon created by the current leadership. Consequently, dramatically different climates may exist simultaneously among the various elements of the Army. The most important determinant of climate is the behavior of leaders. Captain Jordan Becker, who exemplifies adaptive leadership because of his actions in dealing with local communities in Northern Iraq shortly after his unit, the 173rd Parachute Infantry Brigade, combat jumped into Iraq in April 2003, highlights the impact of a positive command climate set by his company commander Captain Jason Ridgeway:

> Jason Ridgeway was an absolutely terrific boss. He gave all of us the leeway to make great things happen in Iraq. Jason Cockman, my good friend and SFAS train-up partner, was a guy who had enormous successes under Jason because Jason (Ridgeway) did what you would've told him to do: *give power to your subordinates*; *teach* them, *mentor* them, and then *trust* them and they will make good things happen for you.[70]

Leaders' behavior directly reflects their perception of people; leadership and management style; skills, knowledge, and attitude; and priorities. Every member of the organization knows that leaders, by their action and inaction, signal what they will or will not tolerate. The leader's behavior creates a climate that influences everyone in the organization. But the overarching culture is determined by the actions of senior leaders and how the members of the Army follow. What happens when small unit leaders try to establish and sustain the climate it takes for their subordinates to succeed (they adapt), while the larger culture does not?

Captain Fagan talked of his experiences in Iraq as a platoon leader of a brigade reconnaissance platoon. He related what his education and training, received prior to being commissioned and supplemented by the Army's Ranger School, did for him as a platoon leader in Iraq, February 2004 through July 2004:

> The ability to make quick, executable decisions with imperfect information was the hallmark of the leadership training that I received at Georgetown. Tactical Decision games, squad and platoon "STX" [situational tactical exercises] lanes, and force on force exercises refined my ability to rapidly assimilate imperfect information, make a decision, and give concise, executable orders to a degree that no other Army school was able to match. Ranger school provided an excellent opportunity to refine these skills and more importantly reinforced the realization that the 70% solution executed violently is always better than waiting to piece together the 100% solution. In addition, Ranger school taught self-discipline, spot checks, teamwork, and leadership in a pressure cooker environment that ROTC will never be able to recreate due to the inherent limitations of a non-combat arms training environment.[71]

Despite this preparation, and the credit that Fagan gave to his noncommissioned officers (NCOs) and how they adapted, there was a constant conflict between the larger culture and the climate that allowed for the adaptation of his platoon:

> The key to my adaptation to combat operations in Iraq lay in my flexibility as a leader and the flexibility of the scouts in my platoon. The oldest man in my platoon was only 30 years old, and the most experienced had only 8 years in the Army, a critical factor in our adaptability. In an Army that still is dead set on training to fight an enemy that no longer exists, the youth that I inherited cannot be overlooked. My men were readily adaptable and were not tied to the book answer on anything. This was critical to our success downrange. [72]

Yet the 2GW culture of the Army did not evolve alongside the climate that Pat Fagan set for his platoon:

> Unfortunately, higher dictated that we label our vehicles, a concession to the insurgents that may have kept us safer, but in my opinion divulges too much information on our identity. In addition, though this war is being fought at the platoon and company level; higher did not give us the freedom to aggressively pursue the insurgents in a manner fitting a 4th GW opponent.[73]

But Fagan's platoon continued to adapt to the 4GW environment, despite the friction created with the 2GW culture. Fagan was proud of his platoon as he pointed to the key trust that was created and sustained, while every member of his platoon came to understand how tactical actions impacted strategy; and reacted accordingly,

> Without the trust, tactical sense, and aggressiveness that we possessed, we would not have been able to enter these two Sadr's Army held cities in only five unarmored HMMWVs, gain and maintain contact, and defend critical coalition assets from repeated assaults of numerically superior enemy forces. Another key factor in our success was our understanding of the 4th Generation Warfare battlefield. Every morning, the platoon began the day with two key priorities of work: maintenance and the *intelligence update*. Every scout in the platoon gathered around a map blow up of the greater Baghdad area and tracked as we read every incident that occurred the day before. Every scout from PVT to SSG was called on to locate the site of the incident, identify the route, assess the enemy TTP, and discuss the history of attacks in that AO. We also discussed the ethnic make up of that AO, the friendly TTPs used or not used in response to the attack, and discussed the other alternative routes in the area that we could use in the event of an ambush, IED [Improvised Explosive Device], car bomb, road block, demonstration etc. My scout PVTs had a better understanding of the battlefield from their morning updates and our wide ranging daily travels than did most S-2s in country, a fact that I credit for our success. I credit our knowledge of enemy TTPs and patterns, the Iraqi culture and pace of daily life, discipline, and agility to react aggressively to any contact with our lack of casualties despite our dangerous mission.[74]

Army senior leaders have the responsibility to set the conditions in a 4GW environment to allow these types of platoons and even squads to succeed. Army generals should be as adaptable as Fagan and his sergeants were expected to be and became in their duties and their approach to accomplishing their missions. If the core competency of an officer like Pat Fagan and his professional NCOs is expertise in the close-in fight and the required preparation for it, the core competency of senior officers is the organization and direction of war according to professional principles within that of 4GW. The ability to adapt was critically lacking in the events surrounding and during this phase of the occupation of Iraq, as Fagan alludes, too:

> My platoon itched at the opportunity to set OPs overwatching known insurgent villages, set ambushes on heavily IED'ed routes, conduct dismounted patrols and ambushes through abandoned military compounds and ASPs. Any of these tactics would have given us the opportunity to catch the insurgents off guard and regain the initiative in our battles with them. Rather than the insurgents dictating the time and place of the battle, we would choose when and where they died. However, time and time again, I was told that we would not be allowed to conduct such operations. When I approached

> higher for permission and laid out plans and areas for our missions I was repeatedly rebuked. Until we make the insurgents fear for their lives and feel like Big Brother may be watching at any moment, they will continue to set IEDs, scavenge abandoned munitions, and build car bombs at will with little fear of retribution. The only language the insurgents understand is brutal, deadly force and we are not willing to take a chance, trust our junior leaders, and really take the war to those executing it with admitted success.[75]

The United States has adapted slowly in response to the new generation of war. 4GW is what American forces encounter in Iraq and Afghanistan and Israelis find in Palestine, and it is the way of the future: evolved guerrilla-style warfare characterized by political acumen and patience, using communications networks and strategic strikes to demoralize and exhaust conventionally superior militaries. For many Army leaders, this new world order amounts to a call to newfangled technological arms, but those trained that using smart bombs and spy drones are the answer got it wrong, and are making the strategic situation worse.

4GW challenges the art of *appreciation* even further than is usually the case in a 2GW or 3GW war between nations, because the non-national character of 4GW combatants embodies a far higher content of squishy political, cultural, and economic factors and relatively less of the traditional military factors. 4GW "terrorists," such as the warriors of Al Qaeda, do not form a conventional army that sets up in military arrays to fight battles, for example. They hide in the cultural and political shadows, and aim to bypass the military to suddenly strike economic, political, and cultural centers of power before melting back into the shadows.

Recent events unfolding in Afghanistan and Iraq raise questions about the efficacy of the America's OODA (observe-orient-decide-act) decision loops in two highly dangerous 4GW conflict situations. The situation in each country is slowly deteriorating, while successes in some command climates have slowed down or even, for brief periods, leveled off the corrosion. Of course this has generated proclamations of victory, like crushing the insurgents. In turn, these drive political statements of *staying the course*. These are not enough to support a coherent OODA-loop, for the simple reason that such statements imply there will never be a need to adapt the assumptions underpinning one's orientation to unfolding conditions. Such statements telegraph weakness.[76]

The Mismatch

In sort of an organizational hypocritical sense, the Army profession has this odd and illogical (irrational) culture of masking our persona (individual and collective) as "the simpleton"—"I'm just a simple kind of guy . . . " is a favored lead-in and credo for many officers in the profession of arms. The Army culture promotes anti-intellectualism (we don't want to be seen as someone who can "think" because that must mean that we can't "act" or "do").

Outwardly, the Army promotes the atrophy of the brain muscle. For telling insight into this, just see what the Army values highly as it evaluates leadership. The impact of a high or maximum score on the Army Physical Fitness Test on "Day-3" of the U.S. Army ROTC Leadership Development Accessions Course (LDAC) usually means that cadet will go on and be in the top of his or her platoon. A failure on the physical fitness test from doing incorrect push-ups has a significant negative impact, destroying any chance to finish in the upper third of his or her platoon or causing him or her to join the largest category for which cadets are "kicked out" of LDAC. This emphasis is instead of building that mental "muscle" the same way soldiers focus on building themselves physically for war. War, it is said, is a young person's game, because of the physical challenges of the business. Or is it?

Warfare is not for the old as defined by out-of-date assumptions; today's 40-something is in as good shape as many 20-somethings. It should also not be left to the weak of mind. The Army appears to appreciate the "thinking-doer" capabilities of its few reformers, while at the same time resenting the very innovators it publicly says it wants. It is an oddity of the organizational culture that must be overcome, and the Army continually demonstrates it has the ability to do so as it tries to adapt "to the demands of counterinsurgency in Iraq at many levels, from the tactical and operational through the training base in the United States."[77]

Many that could make the Army better through their critical thinking have lacked "voice" and access, despite their advanced experience, knowledge, and capacity for thinking, because of their relatively young age and "junior" rank. This is a legacy of the antiquated seniority promotion system. It is also a legacy of the Army's (and larger society's) hyper-focus on "rank."

Leaders will and do adapt to the conditions of war under the "right" command climates. The conditions of war allow good leaders and commanders to put aside and ignore those "engines" that drive the larger culture. "Engines" include the "up or out" promotion system, obsession with quantity over quality as a measure of achievement of "mission," reliance on evaluations that depend on short-term results, an obsession with rank, and an ability to have one's career stymied or even destroyed by one evaluator. Such approaches are used in lieu of strenuous, up-front professional entrance requirements to the officer corps.

So, then what happens when the Army does not adapt as a 3GW culture to 4GW? As the "why" continues to change and evolve, and the Army tries to change everything but its culture, a strategic mismatch occurs with an enemy, shown on the previous pages, that continually evolves to its changing operating environment. A technology-ideas-people versus people-ideas-technology mismatch occurs as seen subsequently. In the past, in previous generations of war, this might have meant a tactical defeat or a setback that was recoverable. Today, setbacks on the battlefield have strategic implications.

When a culture is not evolved in order to set the conditions for the organization, in this case an Army, necessary for its new operating environment, the biggest impact is on its members. Their trust in the top of the organization, their strategic leaders, erodes as rhetoric continues, yet nothing really changes.[78] Then, trust is replaced by frustration, sarcasm, and finally a lack of commitment and acts of self-preservation instead of selfless service. These trends and changes could be observed in the Vietnam War and are chronicled in detail by Dr. Jonathan Shay in his brilliant book *Achilles in Vietnam*.[79]

Part of the cultural problem is hierarchical values. Then it is questionable whether those at the top of the hierarchy will really value something else (human nature). Chris Argyris calls the dichotomy of rhetoric contrasting to reality as "espoused theory," which often differs from the "theory in use." Pinpointing and understanding the difference between these "theories of action" is key to beginning a plan to evolve the culture. A lack of this understanding is what frustrates members of the organization who want and know change is necessary, but over time don't see it occur. They are normally not in a position to impact or influence the change, or feel they do not have a voice in the change.[80]

The Army culture has been learned (both in psychological and social contexts). Hence, culture needs to be unlearned and then relearned. To "re-engineer" culture (many popular management books claim this is possible) to tough because (1) it is not an engineering-type problem and (2) who knows what the culture should be? The members of the organization should be suspicious of those inside the culture (especially "those in power") who define the "right" culture. After all, they are products of the current culture. (How can they be objective?)[81]

Adapting the culture to the demands of 4GW calls for a bottom-up evolution, while those at the top have the strategic vision to "exploit," highlight, and award calls for and innovations that will move cultural evolution. Fortunately, many in and outside the Army have an idea of what the culture should be, and are seeing signs that, from different corners, changes are occurring in reaction to 4GW environments. This happens because junior leaders—officers and NCOs—begin to understand the key to fighting and winning in a 4GW environment is understanding three aspects of war—the physical, the mental, and the moral, with the moral being the most important. This stands in contrast with an "American way of war" that emphasizes the physical aspect of war (destruction of targets).

Other Army officers understand the moral aspect as well and use innovations to solve tactical problems to ensure it remains dominant and does not impact strategy. Lieutenant Colonel Chris Hughes, a battalion commander in the 101st, found his unit faced by an angry mob in Nasiriyah in April 2003 during the U.S. invasion of Iraq. He ordered his people to take off their sunglasses, kneel down and *smile*. It worked! What a way to diffuse a dangerous situation! This was a totally an un-military approach.[82]

But the U.S. Army is getting a better track record of doing that sort of thing, especially inter-agency cooperation or integration. In another insight to the conflict that exists when one level of command practices or attempts to change to Boyd's emphasis on people first, versus the larger organizational culture that demands technological-centric solutions, the author talks about the lack of preparedness of a 2GW military for a 4GW conflict. Lieutenant Colonel Ike Wilson III reports on civil-military planning and operations in northern Iraq, again involving the 101st Airmobile Division, in a paper, "Thinking beyond War: Civil-Military Operational Planning in Northern Iraq."[83]

Lieutenant Colonel Wilson points out U.S. war planners, practitioners, and the civilian leadership conceived of the war far too narrowly—totally misreading Phase IV (occupation) planning. They saw the March–April 2003 invasion of Iraq as the way they wished war would be—quick, cheap—in terms of casualties and deployment costs—versus the reality of war, despite the mounds of evidence that contradicted their fantasy. Wilson goes on to explain how the 101st adapted to stability and support operations in their area of operation, which, because of the adaptations of small unit and some mid-level leaders such as Lieutenant Colonel Hues, went well. Unfortunately, those that followed the 101st were not so adept, and the insurgents regained much of Mosel.[84]

The Army culture has remained focused on the physical aspect of war in everything it does, from placing the centerpiece of evaluation and who gets cut, at Cadet Command's premier leader training, LDAC, on how well a cadet does a push-up, to spending millions of dollars on the future combat system to combat future tanks with high-velocity weapons controlled through a digital, networked, and negative-feedback command-and-control system. It appears now that changes are occurring, mainly from the bottom up, but enabled by the top down.

The Army is just now addressing the mental aspect of war surrounding today's culture from the top and the bottom. This occurs from the top down when the then Army Chief of Staff General Schoomaker pushed officers and NCOs to read books like John Nagl's *Learning to Eat Soup with a Knife: Counterinsurgency Lesson from Malaysia and Vietnam*, or from the bottom up as "lessons learned" by captains who have served in Iraq are passed on to other officers and NCOs on the Internet site companycommand.com.[85]

The bottom-up approach will take the longest and is the best way to change the culture, the type of changes that will take hold and stay. For the first time, articles about Boyd's OODA-loop are appearing in Army professional journals such as *Infantry Magazine*. They are not being written by generals, colonels, or academics, but, better yet, by leaders who have recently applied such theories to the realities of combat, specifically to the 4GW environment. Captain Aaron Bazin writes, "Many company grade infantry officers have probably never heard of Retired Air Force Colonel John Boyd, his way of thinking, or his contribution to the art of warfare."[86]

Bazin's article shows he also had the moral courage to publicly deviate from and question the Army's long-time, established, and methodical doctrine on decision making, the Military Decision-Making Process. Bazin touts how the OODA-loop works well, when understood by the leaders employing it, in the 4GW environment:

> Today's battlefield is constantly changing, based more in urban terrain than ever before, and requires a canny ability to deal with civilian populace, command and control decentralized Soldiers, and fight a tough enemy at the same time.[87]

More significant about Captain Bazin's contribution toward evolving the culture is his use of Boyd's OODA-loop in a 4GW environment, as he continues:

> Faced with new tactical problems set in the environment of Army Transformation, what would an old fighter pilot have to teach us? Simply put, what Boyd can teach you is how the enemy thinks, how you and your Soldiers think, how to train more effectively, and how to control your tactical environment.[88]

Bazin continues to show how he and his soldiers applied the OODA in combat. The article also deviates what is commonly found in Army professional journals; whereas many articles have focused on and proclaimed processes, Bazin's article highlights the Observation and Orient keys in the OODA-loop, with an external focus on the enemy:

> The Boyd decision cycle is a way of looking at how people act in their environment. If a commander can train his Soldiers to minimize their reaction time to tactical problems, train leaders to make sound and timely decisions, and understand and interrupt the enemy's decisions cycle, he gains the advantage.[89]

Additionally, Bazin's conclusion highlights a need to "train leaders to make sound and timely decisions"; but what if these leaders arrived in their units already *trained* to make rapid and timely decisions because they had received the *education* to understand Boyd's OODA-loop? Bazin is one of many junior officers asking for change, and he is trying to drive the change of the Army culture from the bottom up through sharing his ideas through professional journals.

Yet, if the Army is trying a people-centric approach to transformation, it will conflict with DoD and U.S. society, with their obsession with high technology and the physical aspect of war. With little focus on the mental aspect of war, U.S. enemies will continue to out–OODA-loop the United States at the strategic level by having to react to enemy propaganda, threats, and claims.

The solution to winning the global war on terror is by adapting a culture at all levels that understands the power of information, and empowering well-educated and trained "strategic lieutenants" and "strategic corporals" leading and commanding well-trained cohesive units. In turn, everyone understands that moral actions have to be above reproach, or the United States will lose allies, strategic lieutenants will lose local intelligence, and units will lose the ability to operate freely and focusing on a specific enemy. In turn, when all the levels of war lose the moral aspect of war, the nation loses the willing participation of its own people. Evolving the culture to the 3GW where the OODA-loop becomes second nature is necessary for every member to understand, cope with, and win in 4GW. But this cannot occur in isolated parts, just during the times of crisis (in this context, war).

The Army is confusing *climate* with *culture* as it publicly claims when it says the Army culture is changing. In Iraq and Afghanistan, leaders are establishing the right climate for their subordinates to succeed, as the situation with Captain Fagan and the enablers that allowed Captain Bazin to come to understand and use OODA-loop demonstrates.[90] Making such proclamations may be violating one of Kotter's reasons of why strategic leaders fail at changing an organization, declaring an early victory in the quest to change the organization before it is, in fact, accomplished.[91]

Yet, the climate Fagan created was short-lived as he and his key subordinates left the platoon, and they may have even moved into a larger climate that also reflected the larger culture which, in turn, could prevent them from adapting as they did as a platoon in Iraq. Despite their adaptation, their OODA-loop was slowed down by the larger top-down culture founded on an Industrial Age personnel system based on out-of-date assumptions, as pointed out so well by Colonel T. X. Hammes in his book, *The Sling and the Stone*.[92]

At the tactical level, John Boyd believed that complexity (technical, organizational, operational, etc.) causes commanders and subordinates alike to be captured by their own internal dynamics or interactions—hence they cannot adapt to rapidly changing external (or internal) circumstances. At the strategic level, maneuver/countermaneuver suggests we need a potential response for a variety of possibilities: Rapid shifts among many simultaneous and sequential possibilities permit one to repeatedly generate mismatches between the events/efforts an adversary observes or imagines and those to which he must respond (to survive). Without a variety of possibilities, an adversary is given the opportunity to read as well as adapt to events and efforts as they unfold.

Another clue may be found in Boyd's moral war approach. The Army has to constantly generate its own mismatches over the enemy. Rather than micromanagement from above, the answer is a bottoms-up approach, focused to achieve commander's intent. But the overarching culture is determined by the actions of senior leaders and how the members of the Army follow those leaders. An examination of ongoing Army efforts at transformation will reveal whether it is people-centric or remains technology-centric, as subtle changes are made at the edges to protect the status quo.

Conclusion

Americans understand physical war. U.S. technologies help the Army fight and win physical war, and it is where the United States is at its best. Militarily, the Army is beginning to prove that it knows how to fight 4GW in the cultural context, but it remains largely influenced by the larger culture of DoD and U.S. society. First, politicians, the DoD, and the Army must understand it. To understand it requires study and thinking, and the willingness to talk about it openly—about what changes are necessary to win in 4GW.

Translate talk into action in order to evolve the culture. This is the hard part—taking on subjects that were once untouchable, such as the quality of the officer corps and political correctness. The use of political correctness and an outdated personnel system undermines trust as one standard as proclaimed in selections and promotions but, in fact, two standards subtly exist in order to carry agendas beyond what is called for to achieve effectiveness. There appears to be some thinking evolving on the institution of the profession of arms; yet, commissioning sources appear to remain focused upon functioning as officer production lines with more emphasis on numbers as the priority rather than on attaining an understanding of the profession of arms early in an officer's career.

The strategic approach and tactical techniques of 4GW are beginning to force major changes in the way the Army recruits, educates, employs, structures, and trains forces. Professional education, from initial-entry training to war-college level, is being broadened to deal with the wide spectrum of issues commanders will confront in 4GW, while also understanding behaviors required in a 3GW Army. Leaders must be prepared to "group together from a new perspective a number of measures that have been used before but were viewed separately."[93]

Training (as distinct from education) is expanding to deal with tactical situations unique to 4GW, such as staged confrontations between security forces and unarmed women and children. Further, this broadened education must not be limited to military officers but must include decision makers from all agencies of the government involved with international security issues. Parallel to this, the Army must then build a networked, flexible, less hierarchical force structure that employs humans to outsmart the humans fighting and trying to kill us.

The Army is beginning to move beyond joint operations to inter-agency operations. If the enemy is going to strike across the spectrum of human activity, the national response must be coordinated across the multiple national agencies that deal with international issues. Just as joint undertaking had to evolve gradually into full-fledged joint operations, so will interagency operations have to mature from simple meetings to discuss cooperation into fully integrated national operations. Force structure must be reconsidered in light of 4GW issues.

A conflict between rhetoric and reality—the desire to evolve the culture of the Army so it can fight and win in a 4GW conflict with the reality of how hard it is going to be to change the culture founded in the Industrial Age—is why the Army is finding an increasing number of its junior leaders voting with their feet. The "best and the brightest" will be leaving the profession. Despite their abilities to think and act at higher levels of responsibility, they will be ignored because of their rank and grade. Fix this (fix the "trust problem") and the Army will fix the recruitment and retention problem. Then the Army will wind up retaining the "right" folks—those with a "calling to the service" who can prepare and lead the Army in the twenty-first century. With insights into the "why," how is the Army creating the "what"?

Notes

The author is indebted to William S. Lind, LTC Greg Wilcox (USA, ret.), Colonel G. I. Wilson (USMC, ret.), and Colonel Eric Walters (USMC) for their insights contributing to this chapter.

1. United States Marine Corps, *Small Wars Manual of 1940*, reprinted (Washington, DC: U.S. Government Printing Office, 1940), retrieved from http://www.smallwars. quantico. usmc.mil/SWM/1215.pdf.

2. I am indebted to a group of professional Army and Marine (one Royal Marine) officers who meet once a month at William S. Lind's home in Alexandria, Virginia, and discuss warfare, particularly Fourth Generation Warfare. They are some of the best and brightest of the military and include Mr. William S. Lind, Major Andy Dziengeleski (USA), Major Chris Yunker (USMC, ret.), Major David Odom (USMC), CSC Harry Thomsett (British Royal Marines), Colonel Jim Lariviere (USMC), LTC John Sayen (USMCR, ret.), Major John Vanmessel (USMC), Major Todd Simmons (USMC), LTC Mike Golden (USMC), LTC John F. Mcdonough (USMC), and LTC Tim Jackson (USMC).

3. Colonel John Boyd, *Patterns of Conflict*, unpublished briefing (December 1986), retrieved from http://www.d-n-i.net/second_level/boyd_military.htm.

4. My thanks to LTC Frank Hoffman (USMC, ret.) and his thoughts through his unpublished briefing, "Smalls Wars/21st Century" (Quantico, VA: Center for Emerging Threats and Opportunities, Marine Corps Combat Development Command, May 3, 2005).

5. Dr. Antulio J. Echevarria II, "The Problem with Fourth-Generation War," Strategic Studies Institute (Carlisle, PA: U.S. Army War College, February 2005). My issue is only with parts of Dr. Echevarria's interpretation of the "theorists" of Fourth Generation Warfare. The work of Dr. Echevarria on the Prussian Army is, in itself, brilliant.

6. If terminology is what's really important, he may have a point. Any terminology is "artificial," and any metaphor can easily be stretched beyond the point where it provides useful insight. "Species of war" might have been a somewhat better choice of words, since most descriptions of 4GW seem to resemble speciation more than generation.

7. I have heard this even from individuals for whom I have great respect for their leadership and intellect, but their understanding of 4GW is wrong when they say, "Well, it's just another form of insurgency."

8. Echevarria, "The Problem with Fourth-Generation War."

9. William S. Lind, Colonel Keith Nightengale, Captain John F. Schmitt, Colonel Joseph Sutton, LTC Gary I. Wilson, et al., *Marine Corps Gazette* (Quantico, VA: The Marine Corps Association, October 1989), p. 4. Known as Lind, Nightengale, Schmitt, and Wilson from here in.

10. Colonel T. X. Hammes, *The Sling and the Stone: On War in the Twenty-first Century* (St. Paul, MN: Zenith Press, September 2004), pp. 260–266.

11. Ibid.

12. Based on discussions with William S. Lind, December 2004.

13. Discussions via e-mail with Dr. Chet Richards, May 11, 2005.

14. Echevarria, "The Problem with Fourth-Generation War."

15. Ibid.

16. Discussions via e-mail with Dr. Chet Richards, May 11, 2005.

17. Max Boot, *The Savage Wars of Peace: Small Wars and the Rise of American Power* (New York: Basic Books, April 2002), p. 32.

18. I am indebted to Colonel Chet Richards, PhD (USAF, ret.), LTC Greg Wilcox (USA, ret.), and Colonel G. I. Wilson (USMC) for their insights on 4GW.

19. For a description of the first three generations, refer to Chapter 2, "Setting the Conditions."

20. Lind, Nightengale, Schmitt, and Wilson, pp. 4–5.

21. Martin van Creveld, *The Rise and Decline of the State* (Cambridge, UK: Cambridge University Press, 1999), pp. 101–110.

22. Based on discussions with LTC Jan Horvath, *Joint and Multinational Branch, Combined Arms Doctrine Directorate* (Fort Leavenworth, KS, April 13, 2005).

23. Based on discussions with William S. Lind. Also see his columns on military.com.

24. Thanks for the insights of William S. Lind, October 19, 2007.

25. Hoffman, "Smalls Wars/21st Century," p. 6.

26. Discussions with William S. Lind.

27. I am indebted to Colonel Eric Walters (USMC) for walking me through the 4GW opponents, as well as assisting me with the preparation for the course, "The Non-state Soldier," for American Military University.

28. Ralph Peters, *Fighting for the Future: Will America Triumph?* (Mechanicsburg, PA: Stackpole Books, 1999), pp. 4–5.

29. The world's population is growing at 70 million a year. There is a forecast that it will level off at 12 billion by 2050, but while waiting, and hoping this will happen, what is the cost of not doing anything?

30. Jean Raspail, *The Camp of the Saints* (Pestosky, MI: The Social Contract Press, 1987). Donald M. Snow, *UnCivil Wars: International Security and the New Internal Conflicts* (Boulder, CO: Lynne Rienner Publishers Incorporated, 1996).

31. Philippe Delmas, *The Rosy Future of War* (New York: The Free Press, 1995).

32. *World Population Awareness*, retrieved from http://www.overpopulation.org/ on April 15, 2005. "The World Health Organization (WHO) estimates the population to rise to 12 billion people and then stabilize in 2050–60. WHO says this proves that family planning works and all talk about collapse is just cultism - there is no cause for alarm. The WHO projection would make sense if the REST of the earth's living community remained stable. Unfortunately in order to sustain a human population of 6 billion we are losing 70,000 species a year. We are in a period of mass extinction for which the human population is responsible. A human population of 6 billion is not sustainable; the living community cannot indefinitely sustain a loss of 70,000 species a year. As our population grows, the number of extinctions will increase. Our population might become stable at 12 billion but that does not mean the *rest* of the living community would be stable. And our survival depends on its survival. Some experts still have the ridiculous idea that humanity is separate from the rest of the living community."

33. More on this issue is covered in Chapter 3, "U.S. Society's Impact." The United States, which has 6 percent of the world's population, uses 30 percent of the world's energy supply. Twenty percent of the world's population (in other words, its wealthy consumer class) is responsible for over 50 percent of its "greenhouse effect" atmospheric pollutants, 90 percent of its ozone-depleting chlorinated fluorocarbon gases, 96 percent of its radioactive waste . . . and so on.

34. This is one of the arguments of the Wilson or neodoctrine of exporting democracy.

35. Conversations with Colonel Eric Walters, June 2004, former professor for American Military University: "The battle of Agincourt, France, October 25, 1415: The French launched three waves of knights against a thin line of English men-at-arms and

long-bowmen at Agincourt during the Hundred Years War. The French went after what they thought was the threat—the men-at-arms. What killed thousands of the French knights were stakes hidden by the long-bowmen which, when exposed just before contact, stopped the horses and unseated the knights. The flower of French Chivalry paid no attention to the long-bowmen or the possibility of anti-cavalry stakes because they considered both to be cowardly weapons—it takes a real man to fight one-on-one, face to face.

"What is significant about this example is that the French were excellent in execution. They had arguably the best trained knights, armed with the latest types of armor and individual combat weapons, and mounted on magnificent horses. But none of these things mattered when the horses refused to impale themselves into the stakes or were killed by arrows. In fact, they were liabilities as the horses bucked their heavily armored riders into to the ground where Englishmen killed them with halberds.

"If the French had concentrated on finding new threats and worrying about beating them instead of practicing/honing what they already knew how to do best, perhaps the outcome might have been different."

36. Robert D. Steele, *On Intelligence* (Fairfax, VA: AFCEA Press, 2000).

37. James Adams, *The Next World War: Computers Are the Weapons and the Front Line Is Everywhere* (New York: Simon and Schuster, 1998).

38. Walter Gary Sharp, *CyberSpace and the Use of Force* (Falls Church, VA: Aegis Research Corporation, 1999).

39. Colonel Walters breaks down 4GW opponents into four categories: "the high-technology brutes"—North Korea, potentially Russia, China, and Iran; "the low-technology bruts"—warlike ethnic groups, bandits and pirates, narco-traffickers, violent groups with no political agenda; "the high-technology seers"—hackers and hacker spies; and "the low-technology seers"—religious extremists, violent religious/political groups, and terrorists.

40. Colonel T. X. Hammes' talk to Georgetown University Army ROTC corps of cadets, December 5, 2004.

41. Martin van Creveld, *The Transformation of War* (New York: The Free Press, 1991). Peters, *Fighting for the Future*, pp. 23–50. Adda B. Bozeman, ed., *Strategic Intelligence and Statecraft: Selected Essays* (Washington, DC: Pergamon-Brassey's International Defence Publishers, 1992).

42. Hoffman, "Smalls Wars/21st Century," p. 31.

43. Robert D. Kaplan, *The Coming Anarchy* (New York: Random House, 2000).

44. Donald M. Snow, *UnCivil Wars: International Security and the New Internal Conflicts* (Boulder, CO: Lynne Rienner Publishers Incorporated, 1996).

45. Donald J. Hanle, *Terrorism: The Newest Face of Warfare* (Washington, DC: Pergamon-Brassey's International Defence Publishers, 1989). Ralph Peters, *Fighting for the Future: Will America Triumph?* (Mechanicsburg, PA: Stackpole Books, 1999).

46. Admiral Bill Owens and Ed Offley, *Lifting the Fog of War* (New York: Farrar, Straus, Giroux, 2000).

47. General Richard Myers, chairman of the Joint Chiefs of Staff, speech (September 2004).

48. David Callahan, *Unwinnable Wars: American Power and Ethnic Conflict* (New York: Hill and Wang, 1997).

49. Gregory Cochrain, "Bush's Napoleon Complex: What the French Experience in Spain could teach us about Iraq," *American Conservative* (Washington, DC: March 28, 2005), retrieved from http://www.amconmag.com/2005_03_28/article.html. "Napoleon's

army in Spain ended up controlling only the ground it stood on. The roads weren't safe—every supply convoy needed an armed escort. The struggle against guerrillas was never-ending. The French, who had thought of themselves as bringing enlightenment, ended up hating the Spanish. This all sounds terribly familiar, but the parallels do end. France lost, but the U.S. won't. Spain was weaker than France but not militarily insignificant, and it had Great Britain backing it with money, troops, and Wellington. We're hundreds of times stronger than Iraq. The U.S. may tire of a pointless war and leave, but we certainly won't lose battles."

50. Interview with Colonel Chet Richards, PhD (USAF, ret.), January 12, 2005; "We are able to show the evolutions of war across a spectrum of time." Also thanks to William S. Lind and Colonel T. X. Hammes for exposing me to a different perspective regarding the generations of war.

51. General Kevin Byrnes, remarks at Association of the United States Army (AUSA) Institute of Land Warfare Breakfast (Arlington, VA: March 3, 2005).

52. Franklin C. Spinney, "Evolutionary Epistemology: A Personal Interpretation of John Boyd's Destruction and Creation," unpublished briefing (Washington, DC: September 1997). Bruce Gudmundsson, "Maneuver Warfare: The German Tradition," in *Maneuver Warfare: An Anthology*, ed. Richard D. Hooker Jr. (Novata, CA: Presido Press, 1993).

53. G. I. Wilson, Greg Wilcox, and Chet Richards, "Fourth Generation Warfare and OODA Loop Implications of the Iraqi Insurgency," *Defense and National Interest* (December 22, 2004).

54. I appreciate the insights of Major Andrew Dziengeleski, Mr. Robert Krumm, Mr. Mark Lewis, Major Chris Coglienese, Major Darryl Perry, and Major Ike Wilson.

55. Lind, Nightengale, Schmitt, and Wilson.

56. Dr. Chet Richards, *A Swift, Elusive Sword: What if Sun Tzu and John Boyd Did a National Defense Review* (Center of Defense Information, May 2001).

57. Brigadier General Huba Wass de Czege (USA, ret.) and Major Jacob D. Biever, "Soldiers – Not Technology – Are the Key to Continued Superiority," *Army* (Arlington, VA: Association of the United States Army, March 2001), p. 1.

58. Frederick W. Kagan, "A Dangerous Transformation," *The Opinion Journal,* WSJ.com (November 12, 2004), retrieved June 18, 2004, from http://www.opinionjournal.com/forms/printThis.html?id=110004289.

59. Mark Lewis, "Time to Regenerate a GenX Response to Dr. Wong's Monograph," *Defense and National Interest* (November 2000).

60. Kagan, "A Dangerous Transformation"

61. Dr. Leonard L. Lira, "To Change an Army: Understanding Defense Transformation," paper for presentation at the 2004 ISSS/ISAC Annual Conference (West Point, NY: Department of Social Sciences, October 2004), pp. 23–24.

62. Lira, "To Change an Army: Understanding Defense Transformation," p. 25.

63. Ibid., p. 25.

64. George Reed, Craig Bullis, Ruth Collins, and Christopher Paparone, "Mapping the Route of Leadership Education: Caution Ahead," *Parameters* (Carlisle, PA: U.S. Army War College, Fall 1994).

65. Online interview with First Lieutenant Patrick Fagan (USA), April 13, 2005.

66. George B. Forsythe, Scott Snook, Philip Lewis, and Paul T. Bartone, "Making Sense of Officership: Developing a Professional Identity for 21st Century Army Officers," in Snyder and Watkins, *The Future of the Army Profession* (New York: Custom Publishing, 2001).

67. Williamson Murray, "Armored Warfare: The British, French and German Experiences," in *Innovation in the Interwar Period*, Williamson Murray and Allan R. Millett, eds., (Cambridge: Cambridge University Press, 1996), p. 122.

68. Leonard Wong, "Stifled Innovation? Developing Tomorrow's Leaders Today," *Strategic Studies Institute* (Carlisle, PA: U.S. Army War College, April 2002), pp. 1–22.

69. Lind, Nightengale, Schmitt, and Wilson.

70. E-mail interview with Captain Jordan Becker, May 3, 2005.

71. Interview with Fagan.

72. Ibid.

73. Ibid.

74. Ibid.

75. Ibid.

76. Reference the columns of William S. Lind on http://www.military.com. Also based on numerous discussions with Mr. Lind, LTC Greg Wilcox (USA, ret.), Colonel G. I. Wilson (USMC), and Colonel Chet Richards, PhD (USAF, ret.).

77. LTC John Nagl, "Spilling Soup on Myself," preface to *Learning to Eat Soup with a Knife: Counterinsurgency Lessons from Malaya and Vietnam*, 4th ed. (Chicago, IL: Chicago Publishing, 2005).

78. Kotter references this in his eight errors an organization makes when trying to change as declaring victory too early.

79. Dr. Jonathan Shay, *Achilles in Vietnam* (New York: Simon & Schuster, 1995), pp. 23–24.

80. Dr. Chris Argysis, "Making the Undiscussable and Its Undiscussability Discussable," *Public Administration Review* (Boston, MA: American Society for Public Administration, May–June, 1980), pp. 205–213.

81. E-mail discussions with Colonel Chris Paparone, former instructor on leadership, U.S. Army War College.

82. LTC Chris Hugh's talk to Georgetown Corps of Cadets, April 15, 2005.

83. Major Isiah Wilson, March 12, 2005.

84. Discussions with Major Isiah Wilson, December 2004 through April 2005.

85. Companycommand.army.mil, "Company Command—Building Combat-Ready Teams: Reflexive Fire Training—Taking Marksmanship to a New Level," *Army* (Arlington, VA: Association of the United States Army, May 2005), p. 13; also retrieved from http://www.ausa.org/, May 2, 2005.

86. Captain Aaron A. Baxin, "Boyd's O-O-D-A Loop and the Infantry Company Commander," *Infantry* (Fort Benning, GA: U.S. Army Infantry Association, January–February 2005), p. 17.

87. Ibid., p. 17.

88. Ibid., p. 17.

89. Ibid., p. 19.

90. Colonel Richard H. Mackey, Sr., "Translating Vision into Reality: The Role of the Strategic Leader. Student Study Project" (Carlisle, PA: U.S. Army War College, April 15, 1992). Edgar H. Schein, *Organizational Culture and Leadership* (San Francisco, CA: Jossey-Bass, 1992). John K. Setear, Carl H. Builder, M. D. Baccus, and Wayne Madewell, *The Army in a Changing World: The Role of Organizational Vision* (Santa Monica, CA: The Rand Corporation, June 1990). Don M. Snider, *The National Security Strategy: Documenting Strategic Vision* (Carlisle, PA: Strategic Studies Institute, U.S. Army War College, February 24, 1992).

91. John P. Kotter, "Winning at Change," *Leader to Leader* 10 (Fall 1998), pp. 27–33.

92. Hammes, *The Sling and the Stone*, pp. 271–290. Many of Colonel Hammes' other recommendations will be familiar to those who have read *The Path to Victory*, which Hammes credits as the basis for his own personnel proposals: Solve the people problems and our troops will figure out ways to employ suitable technologies. Hammes' application of the ideas in *The Path to Victory* to fashioning a military capable of 4GW is among the most innovative parts of the book and potentially among the most decisive.

93. Discussions with Colonel T. X. Hammes, December 2004.

CHAPTER 2

History of Manning

> "As soon as public service ceases to be the main business of the citizens, and they prefer to serve with their pocket books rather than with their persons, the State is already close to ruin."
>
> Jean Jacques Rousseau[1]

In the nineteenth century, the United States relied on state units (evolved into what is known today as the National Guard) possessing little military experience for its wars and most particularly for the Civil War. This shaped military thinking for the next half-century, and largely influenced today's military cultural foundation—the U.S. mobilization doctrine. Unlike the South, which immediately promoted former regulars to command its units, governors in the North appointed cronies and lost the first battles. Later, when units became battle-experienced but casualty-depleted, governors again favored cronies over experience and organized new units with new colonels rather than replenishing their already existing units. The South also, on the other hand, continued to replace losses in their experienced units with personnel from the same region as the original unit was raised until it completely ran out of manpower in the summer of 1864.

Nonetheless, the Civil War replacement system—for all of its shortcomings—at least had the virtue of regimental recruiting from the same locality (increasing bonding likelihood), and regiments (really large battalions by today's standards) remained cohesive and fought well even with effectives as low as 30 percent of authorized strength.

The regular Army was resentful and determined that it not be promoted over again. The obvious solution was to structure new units on mobilization around rapidly promoted regulars in the high command and staff positions and to flesh the remainder with newcomers. When demands for casualty replacements zoomed beyond all expectations 1917, state units were bled and their command structure relegated to training. This solution—individual "dribble" replacement—was conveniently justified as the latest in industrial efficiency, as proposed by Frederick Taylor and his managerial disciples.[2]

Experience was gained—but cohesion suffered in the European Theater in both World Wars I and II. In the Pacific Theater, the Marine Corps escaped the perils of individual replacement, as did the Army, which had a psychiatric rate higher than any of the other Western militaries in Europe but lower than any of theirs in the Pacific—revealing much about the problem and its cure.[3]

In 1947, thinking to correct manning deficiencies that had become so apparent in the 1917 and 1940–41 mobilizations, cadre were further increased and "up and out" promotion adopted to reduce perceived problems of a stagnating and aging officer and noncommissioned officer corps. This manning system envisaged a war with the communist powers in the not-distant future. With time, it has grown into a misconceived, self-serving solution. It has failed to incorporate the better solutions available for wartime mobilization. Among its unintended consequences are peacetime careerism and too many field grade officers.

General Paul Gorman's (USA, ret.) excellent institutional history, *The Secret of Future Victories,* credits George C. Marshall as the architect of the modern version of the current manning system. He was personally involved in senior staffing during the war, and in correcting perceived shortfalls afterward.[4] He believed the success of the American Army in World War II was due to divisions that were brought "up to strength daily by trained men from the replacement pool."[5] The Marine Corps also followed this belief, although the peculiarities of short island campaigns often precluded actually implementing it.

The driving factor in General Marshall's (as well as MacArthur's) thinking in structuring an expanded Army was the experience factor: A small regular officer corps could not be spread too thinly.[6] From the perspective that divisions were linearly dispersed firepower-grinding machines and casualties were focused on the relatively small infantry component, Marshall's thinking maximized the use of the peacetime officer corps and was cost-effective in the use of the putatively more difficult to train technical specialties. Infantry paid the price.

Another result was the limited availability of reserves for replacing, resting, and re-training tired units. For a given manpower strength, the United States could only field 89 Army and 6 Marine divisions. Other armies with different force structuring and manning practices fielded three to four times more divisions with equal "bayonet" strength and considerably more automatic weaponry, and fought a protracted war over a long and poorly endowed line of communications. Why the low number of divisions for the United States? The linear format presumed by General Marshall, and everyone else except the Germans and the Tukhachevski wing of the Red Army, assigned combat support and combat service support organically instead of using the *Schwerpunkt* (main effort) principle-like focused logistics, as the Army has termed it.[7]

Mobilization Doctrine Foundation

Today's personnel management paradigm—how the Army mans and sustains its legions—evolved from one that worked to support the nation's long-standing mobilization doctrine. Mobilization doctrine relies on a small core of full-time professional soldiers, backed by large militias or National Guard forces in peacetime, to be prepared to expand rapidly in the event of a national emergency such as war. Successful mobilization requires time and massive resources. Time is needed to prepare troops, while resources compensate for lack of experience, professionalism, and cohesion needed to fight and win a war.

To support the mobilization doctrine, the Army developed recruiting and training methods that paralleled management practices in the corporate structures of the Industrial Age. The challenge for the Army was to get millions of citizens with little or no military experience and turn them into soldiers and officers in a short time. Industry provided the answers, and in the aftermath of the glow of victory in several wars, these approaches became institutionalized.

Some modifications have been applied to the personnel management paradigm, but they happened along the fringes of existing laws, regulations, policies, and beliefs. Army alterations to today's leader development paradigm may not be enough. The Army has "thought" and "acted" from an antiquated, mobilization-based leader development paradigm that still exists more than 19 years after the fall of the Berlin Wall. This Industrial Age model continues to shape the way the Army approaches its recruiting, training, and education and how it mans its legions. Consequently, the Army's personnel development paradigm—designed for an earlier era—has been so intimately tied to the maintenance of current Army culture that a self-perpetuating cycle has formed that diminishes and, in some cases, even prevents the Army from developing adaptive leaders and units to deal effectively with Fourth Generation Warfare (4GW).

Manning the Force for the Past

One cannot examine how our nation mans its legions without understanding the relationship with the history of the draft (conscription) and volunteerism. Though the draft or conscription occurred only during 35 years of the Army's history, it was the centerpiece in how to man the Army prior to the introduction of the all volunteer military in 1973. When recruiting is examined in the history of the U.S. Army, it is about how to "man" the forces with available manpower. Prior to this, conscription was primarily used to man America's legions in times of war while units were sustained using what is called the Individual Replacement System.[8]

Recruiting for the U.S. Army began in 1775 with the raising and training of Continentals to fight in the Revolutionary War. The U.S. Army Recruiting Command, however, traces its history back to 1822, when the General Recruiting Service was started by Major General Jacob Jennings Brown, commanding general of the Army. Recruiting would be difficult in a new nation that distrusted standing armies. There were good reasons for this. Americans "remembered the British Army's lifetime enlistments, the press gangs that seized vagrants off the streets, the judges who released convicts on the condition that they would join the Army, and other abuses commonly practiced to obtain men for military service."[9]

Recruiting from the Revolution up to World War I followed a familiar pattern. Recruiting parties were sent from each regiment, with the regimental colonels being admonished to select "active and vigilant recruiters who stood high in the esteem of the people in the districts in which they were to solicit." Recruiters were usually officers, who would travel to "places of rendezvous where recruits assembled." These officers would represent their regiments and were recruiting to keep their units filled. These recruiters were not recruiters in the traditional sense as we know them today. Recruits assembled at the rendezvous points at set dates, often traveling many miles to do so, at which point the recruiter showed up to conduct physical examinations, and those recruits who passed then took a sworn statement before a peace officer, a general officer, a judge advocate, or deputy judge advocate. The recruit then signed his enlistment papers, and was marched away with the other recruits under the direction of the recruiting officer to begin their military training immediately.[10]

For our forefathers, conscription was a normal way to supply an army with manpower. During the war, a quota system was used to man the Continental Army. The writers of our Constitution addressed Congress's authority on this matter in Article I, giving Congress the power "to raise and support Armies" and "to provide for calling forth the Militia to execute the Laws of the Union, suppress Insurrections and repel Invasions." In times of crisis, the states would call upon their militias to raise a required amount of manpower, and then integrate into Federal service.[11]

The Constitution gave Congress more power over the military than the Articles of Federation had, and created some debate. However, the discussion was not about how the armies would be manned. The importance of this debate had less to do with compulsory service and more to do with the distinction between standing armies and citizen militia armies. A drill manual published shortly before the Revolution stated that militias are composed of men who "thought for themselves" while standing armies were made up of men who only served for their pay. The Constitution did not spell out whether conscripting soldiers was within the scope of power of the U.S. government. Up to the Civil War, the pattern of recruiting, as described earlier, was adhered to; with the nation's first modern war, other ways to man the nation's legions had to be found.[12]

The Civil War

The Civil War was the first war in which the United States mobilized its entire population in order to carry out its national objective, which was to reunite the seceded Southern States with the North. The militia system established out of mistrust for standing armies was soon inadequate for providing the hundreds of thousands of soldiers needed to fight along a front of more than one thousand miles. Within the first year of the war, both North and South had to revert to conscription as the harsh reality of modern war diminished and eliminated the rush and flood of the initial volunteers.

The South was the first side forced by conditions to choose conscription over volunteerism. General Robert E. Lee recommended to President Jefferson Davis that a conscription bill be enacted. On April 16, 1862, the Confederate Congress enacted the first national draft law passed by an American legislative body by a vote of more than two to one. The laws provided that all white males between the ages of 18 and 35 were liable for military service for three years. Later in the war, Congress extended the age limits from 17 to 50. The draft raised only about one-third of the Confederacy's military power. Several problems weakened the draft's ability to raise more soldiers. Governor Joseph E. Brown of Georgia called the law "a palpable assault upon the rights and sovereignty of the State" and listed 15,000 persons as indispensable state officials who were not liable to the draft.[13]

In 1862, the Union also confronted similar manpower shortages. The Militia Act of 1862 had a clause that provided President Abraham Lincoln with the authority to draft manpower from the states that did not meet their quota of volunteers, representing the first time in the history of the United States that the Federal Government assumed military conscription power. However, the draft of 1862 never went into effect, as this threat of conscription had the effect of spurring enlistments by volunteers.[14]

On March 3, after two weeks of debate, Congress passed by comfortable majorities the Enrollment Act of 1863 (also known as the Conscription Act). Under the constitutional clause permitting the government "to raise and support armies," the Act imposed a draft liability on all male citizens between the ages of 20 and 45 years for three years or the war, whichever ended first. Like the Confederate conscription law, this Act permitted substitution and commutation, whereby a draftee could purchase exemption for $300.[15]

Of the approximately 2,670,000 men raised by the North, the direct effect of the draft can account for only 6 percent of this total. The draft's major purpose was to spur enlistments by threatening conscription. States expended huge sums on bounties. The Civil War marked the last major conflict for which the nation would depend on both organized militia and volunteers. Professor Weigley summarized his perspective on the war's manpower policies:

> After reviewing the problems and anomalies of the effort to maintain the volunteer armies of the 1860's, the United States would never again attempt to raise a mass wartime army by that method. Federal conscription would be the principal legacy of the Civil War experience to future American war armies.[16]

Additionally, an after-action report by Brevet Brigadier General James Oakes, the assistant provost marshal general who administered the draft laws in Illinois, recommended that "it be the citizen's responsibility to register in future drafts; that substitution, commutation, and bounties be rejected forever; and that the central government take over full responsibility for the draft rather than depend on the states for help." This report would play a significant role in shaping the draft during World War I, when his recommendations were implemented.[17]

World War I

President Woodrow Wilson called for America's distance from the European war and believed that volunteers would be sufficient to meet the nation's military needs in his State of the Union message to Congress on December 8, 1914. Members of Congress stood cheering when Wilson finished his address. Two years later, Congress passed the Selective Service Act of 1917 by overwhelming majorities.[18]

The Preparedness Movement leading to the Selective Service Act gained momentum under the direction of former President Theodore Roosevelt and former Army Chief of Staff Major General Leonard Wood in 1915. The American Rights Committee, the National Security League, the Army League, and the National Association for Universal Military Training were some of the influential interest groups of this movement. General Wood was the guiding spirit of the Plattsburg movement, which was a voluntary military training program that many prominent business and professional men attended. He argued that "national preparedness means . . . first of all, the moral organization of the people, an organization which creates in the heart of every citizen a sense of his obligation for service to the nation in time of war."[19]

The major push for conscription came from the Army's General Staff and its War College Division. In November 1915, the Secretary of War, Lindley Garrison, forbade consideration of the topic when he learned that a General Staff captain was circulating a memorandum advocating compulsory military service. Major General Hugh L. Scott, Army chief of staff, and Secretary of War Newton D. Baker, who succeeded Garrison in March 1916, were the two dominant players who would change the president's position on conscription, guide its passage through Congress, and ensure its effective implementation.[20]

In hearings before the House Military Affairs Committee on January 10, 1916, General Scott advocated compulsory military service, but emphasized that "I can speak only for myself." His belief, which was contrary to the president and secretary of war's positions, was a bold public stand for the senior military man at the time. That same month, General Scott directed the War College Division to prepare a study of a system of training and universal service to form the basis of legislation. General Scott approved this study, called the National Army Plan, on February 14, 1917, and Secretary Baker submitted it to Congress on February 23.[21]

After the war, Secretary Baker described the manner in which he and the president became proponents of conscription: "After the suspension of diplomatic relations between the United States and Germany [on February 3, 1917], General Scott, discussing with me the possibility of our entrance into the war, raised the question of the method by which men should be called into service." The passage illustrates that President Wilson made the decision to conscript America before he called on Congress to declare war and more than two months before Congress passed the Act that legalized the draft. By closely held agreement, the president, secretary of war, and Major General Enoch Crowder, judge advocate general of the Army, organized "the colossal machinery for enforcing the draft . . . long before the country knew there would be a draft—while, indeed, the country continued to take it for granted that only the volunteer system would be used."[22]

President Wilson approved conscription for two reasons. First, he recognized the efficiency and equity of the draft over the chaos of the volunteer system. Second, he knew that by opting for conscription, he could block one of his leading critics, former President Theodore Roosevelt, who wanted to raise a volunteer force and lead it in France. Now with the president's support, General Crowder drafted the legislation along with Secretary Baker, Major General Scott, Major General Tasker Bliss, and Brigadier General Joseph Kuhn during a series of conferences. The following day, the draft bill was submitted to Congress with President Wilson's endorsement for "men, who should, in my opinion be chosen upon the principle of universal liability to service."[23]

Both houses of Congress engaged in heated debate over the draft legislation for over a month. Senators in favor of selective service argued that "volunteering is haphazard, inefficient, disruptive of industrial and economic stability, wasteful, and operatively unequal in spreading the obligations of citizenship" and "conscription is the fairest, most democratic, most efficient and most patriotic method of raising an army."[24]

The results of a survey of American newspapers and mayors by a preparedness interest group, the National Security League, indicate why overwhelming majorities in both houses supported its passage, reflecting a barometer of the American public's feelings on the draft. President Wilson stated, "It is in no sense a conscription of the unwilling; it is rather, selection from a nation which has volunteered en mass." However, it was not entirely en mass as more than 300,000 men evaded the call to service.[25]

America's use of selective service provided a model for the intelligent use of manpower in wartime. The draft supplied 67 percent of the manpower needed for the war. On registration day, President Wilson told his fellow Americans that while "the nation needs all men," it needs them where they "will best serve the common good. . . . The whole nation must be a team in which each man shall play the part for which he is best fitted."[26]

Echoing his commander-in-chief, General Crowder stressed that "the ultimate goal of America was to organize not only an army, but a nation for war. . . . Every man within the draft age had to become either an effective producer or a soldier."

Allan Millett and Peter Maslowski, military historians, wrote: "As intended, the draft 'selected' those men the Army wanted and society could best spare: 90 percent of the draftees were unmarried, and 70 percent were farm hands or manual hands." The Selective Service Act of 1917 was strictly a wartime measure; one of its provisions explicitly stated that conscription would end with hostilities.

General Crowder, who drafted the legislation, stressed the key to Allied victory: "Just as the war could not have been won without America, so the war could not have been won without the draft."[27] After the war, Congress considered universal military training, but there was very little popular support for compulsory peacetime military service.

World War II

Despite the widely held belief that the Army of WWII was manned by volunteers, it was conscription, or the draft, that enabled the nation to man an Army with 8.3 million personnel while fielding 89 divisions and numerous independent units such as corps and army support units, and independent tank and artillery battalions. The system had the nation's support because it was a global crusade against an evil enemy, but the United States also understood "who mobilizes the elites, mobilizes the public." The efforts of the National Emergency Committee (NEC) of the Military Training Corps Association and their leader, Greenville Clark, a prominent lawyer and anglophile, remain among the true behind-the-scenes heroes of WWII. Clark became known as the "Father of Selective Service."[28]

The White House, War Department, and the Army sat on the sidelines and Germany continued its attack against France and Great Britain, while "Clark and his organization, in four short months, conceived, wrote, and lobbied the Selective Training and Service Act (STASA) of 1940 through Congress during an election year. Led by Clark, the NEC decided to institute a peacetime draft and strengthen the Army. President Roosevelt did not want the political liability of supporting a peacetime draft during an election year, especially his own unprecedented third term."[29] Roosevelt shared "doubt as to whether a limited form of selective draft will be popular. When the Republican presidential nominee, Wendell Willkie, stated that "some sort of selective service is the only democratic way in which to secure the competent and trained manpower we need for national defense," the draft was essentially removed as an election issue. Finally, on the verge of war, both Roosevelt and Willkie accepted the need for conscription, but Congress took three months of debate until finally passing the STASA in June 1940.[30]

STASA was effective in raising the manpower necessary for the United States to wage World War II. Nearly 50 million men registered and 10 million were inducted. Selective Service supplied two-thirds of the manpower for the U.S. military during the war and enabled the Army to reach its peak strength of 8.3 million men in 1945. Five million draftees served in Europe, Asia, Africa, and the Pacific.[31]

At the same time, "STASA also granted occupational deferments to five million workers to ensure that critical labor was done at home in support of the war."[32] The STASA system was well run; survey after survey during the war showed unprecedented public approval with the operations of the draft, leading George Gallup to remark, "Few programs in the nation's history have ever received such an overwhelming favorable vote."[33]

The Cold War

Despite the public's and politicians' desire that World War II would end the need for a large military, particularly a large Army, new threats arose that dictated otherwise. During the postwar years, the Army faced a continual shortage of manpower. George C. Marshall warned the Truman administration that "we are playing with fire while we have nothing with which to put it out."[34] Though STASA was extended a couple of years after the war, it ended on March 31, 1947, and for a year the military relied on volunteers. The Army needed 30,000 volunteers a month at that time, but only 12,000 were enlisting. General Dwight Eisenhower retired in February 1948 and warned the administration that the Army was 100,000 below strength, and he also anticipated greater shortages.[35]

In response to the growing threat from the Soviet Union, as well as numerous warnings from several prominent World War II general officers, the Truman administration pushed for and received the Selective Service Act of 1948. This Act provided for 21 months of active Federal service with all men from ages 18 to 26 required to register. While not inducting a lot of Americans, the Act did encourage voluntary enlistments of people who wanted to have a choice where they served. This Act filled the ranks of the military with volunteers and, as Professor Eliot Cohen wrote, "the draft had, as during the Civil War, provided large numbers of men by eliciting eight or ten volunteers for every conscript."[36]

The Korean War in June of 1950 forced Congress to extend the Act of 1948, but then after the first year of the war, Congress revised the earlier Act through the passing of the Universal Military Training and Service Act of 1951. This extended the president's authority to induct citizens for four years, granted him the authority to recall reservists, lowered the draft age to 18, lengthened the term of service to two years, and cancelled deferments for married men without children.[37]

The draft continued throughout the 1950s, even as the Korean War slowed down by 1953 with the signing of the armistice. Afterward, the draft was considered to be "wasteful of young talent," and the short-term two-year service was costly to both the individual and the DoD.[38] From the 1950s through the 1960s, DoD continued to argue and receive support for a draft by advocating five points:

- "the draft simulated enlistment in all Services";
- "higher aptitude candidates were induced by draft pressure to voluntarily enlist in 3- and 4-year training programs";

- "the threat of being drafted supported officer training programs";
- "conscription supported the reserve forces";
- and "the all volunteer policy failed from 1947 to 1948 and would be too expensive to sustain and equip the force size required."[39]

Through the later years of the Vietnam War, these views remained popular and supported by the U.S. public. The draft was repeatedly renewed by overall majorities in Congress in 1955, 1959, and 1963. In 1967, the final extension of the draft was also passed by a majority of Congress after a year of hearings and public debate.

For 27 years, from 1945 to 1973 (with the exception of one year, 1947–48), the United States manned its legions with conscription. From 1965 to 1973, there were 1,728,254 inductions through selective service. As opposition and resistance to the protracted conflict in Vietnam intensified and grew, so did protest against the draft system and these inductions. In 1956, 77 percent of the population supported the draft. In 1970, a bare minority supported the draft; one survey in 1972 reported only 13 percent did. With a shift in public support this large, it is perhaps easy to see why President Nixon led the charge against the draft and advocated volunteer recruitment.[40]

Volunteers Again Fill the Ranks

When President Lyndon Johnson thought about sending reinforcements to Vietnam in 1965, he sought former President Eisenhower's opinion. Eisenhower approved of the increase, but warned that "sending conscripted troops to Vietnam would cause a major public-relations problem," which is actually what happened to the U.S. government and its politicians. Robert Griffith, a historian, provided a sound explanation why "the draft was a natural casualty of the longest, most unpopular war in American history":

> Inductions from 1954 to 1964 averaged 100,000 a year. As American involvement in Vietnam escalated, so did conscription... In 1966, 400,000 were called. Casualties also increased, especially among draftees. Draftees, who constituted only 16 percent of the armed forces, but 88 percent of infantry soldiers in Vietnam, accounted for over 50 percent of combat deaths in 1969, a peak year for casualties. Little wonder that the draft became the focus of anti-Vietnam activism.[41]

Because of the various exemptions granted to students pursuing college, graduate, and professional degrees, high school dropouts were twice as likely to be drafted as college graduates. Protest against the war focused on the draft system that was "sending minorities and underprivileged youth to die in the jungles of Southeast Asia."[42]

Because of the growing resistance to manning through a draft, President Richard Nixon contemplated a military manned by volunteers. Nixon "decided to end the draft not because it was failing but because its political cost had become too high." By attacking the draft, Nixon would gain "a positive image, especially on the campus and with intellectuals generally"—two groups normally supportive of the Democratic Party.[43] On October 17, 1968, Nixon, the Republican presidential candidate, made a campaign promise to seek to end the draft when the Vietnam War was over. In his campaign speech, Nixon stated:

> A system of compulsory service that arbitrarily selects some and not others simply cannot be squared with our whole concept of liberty, justice and equality under the law . . . the only way to stop the inequities is to stop using the system. . . . Today all across our country we face a crisis of confidence. Nowhere is it more acute than among our young people. They recognize the draft as an infringement on their liberty, which it is. To them, it represents a government insensitive to their rights, a government callous to their status as free men. They ask for justice, and they deserve it. So I say, it's time we look at our consciences. Let's show our commitment to freedom by preparing to assure our young people theirs.[44]

Another major influence on Nixon and his stance on the All Volunteer Force (AVF) was Dr. Martin Anderson, "the person who was almost singly responsible for placing this policy issue in front of a policy-maker and seeing that the policy decision was implemented."[45] Anderson was a young economics professor who worked on the president's campaign, then was a senior White House advisor. Anderson repeatedly told Nixon that the AVF was an issue he could "use to establish a rapport with the youth of the country."[46] Anderson made key recommendations regarding the composition of the president's blue ribbon panel to study the AVF and he worked feverishly with various economists to justify the economics of the AVF.

With this background, President Nixon appointed a commission, led by former Secretary of Defense Thomas Gates, "to develop a comprehensive plan for eliminating conscription and moving toward an All Volunteer Armed Force." On May 13, 1969, even before this commission submitted its report, President Nixon informed Congress that reform meant replacement with volunteers in his Special Message to Congress on Reforming the Military Draft.[47]

In February 1970, the Gates Commission released its favorable AVF report: "We unanimously believe that the nation's interests will be better served by an all-volunteer force, supported by an effective stand-by draft, than by a mixed force of volunteers and conscripts; that steps should be taken promptly to move in this direction." The Gates Commission's basic premise was that "conscription is a form of taxation, the power to conscript is the power to tax." Military service was no longer seen as an obligation, but as a tax on America's youth. The Gates Report further stated: "When not all our citizens can serve, and when only a small minority are needed, a voluntary decision to serve is the best answer, morally and practically, to the question of who should serve."[48] The Commission believed that the military could entice enough volunteers to enlist by increasing pay, improving conditions of service, and more vigorous recruiting.

The United States implemented a remarkable change in its manpower policy in 1973. Prior to the creation of the AVF, the United States had a draft for 27 years during the Cold War, in which a composite force of draftees and volunteers served. It was widely accepted by politicians and the American people for all of that period except for the last three to four years, when Vietnam protests became intertwined with draft protests. "There were two primary reasons for this broad public acceptance for conscription: the nature of the Communist threat with the Soviet Union and the need for a reliable and steady source of manpower which the draft provided and volunteers did not."[49]

Conclusion: Draft or Volunteerism?

The only justification for a draft is when there is a national emergency confronting the country. Thus, external threat is a key factor, present during every draft period. Not only was the threat real, but it was also one that was not ending anytime soon. Although the external threat and the resulting perceived long-term emergency were important and necessary conditions before each draft enactment, these alone were not sufficient.

For the Civil War, the problem for the Union was the Confederacy's secession. In both World War I and II, it was primarily Germany. The Cold War problem was the Soviet Union and Communism. But, what is it today? These problems "opened" a policy window to allow political actors—presidents, members of Congress, the military, influential citizens, and lobbying groups—a chance to advocate, push, and enact their policies or to block those of others. During these draft periods, the actors saw volunteering as a policy that was unable to meet the nation's manpower needs.[50]

For military issues that involve raising and supporting an army, the commonly held assumption is that Congress plays the dominant role in enacting the policy. This is not always the case though, especially given the analysis of the four major draft periods. Perhaps because of the nature of the U.S. political system with its pluralistic society and the multiple points of access provided by our government's separation of powers, it should not be a surprise that each draft period had a different dominant actor.

During the Civil War, Congress took the lead, although President Lincoln and Secretary of War Stanton operated outside the public's view. In World War I, the dominant actor was the military, especially General Scott and the Army General Staff, which pressured first the secretary of war, and then President Wilson to act. Prior to World War II, it was Greenville Clark, a prominent, rich, and well-connected attorney, and his group that pressured Congress to act while President Roosevelt, General Marshall, and the War Department were hesitant and nonsupportive. And during the Cold War, the dominant actor was the president, who called for draft extensions and kept peacetime conscription as the permanent manpower policy for the United States with overwhelming support by Congress. Likewise for the creation of the AVF, President Nixon was the dominant actor who responded to the domestic political crisis that resulted from the Vietnam War.

There are five insights from this short analysis of the history of how the United States has manned its legions.

- The Industrial Age management revolution of Frederick Taylor has had a profound impact today on how the U.S. military, particularly the Army and Marines, man and sustain their units.
- The draft was used only during 35 years of our nation's history. Volunteering did not produce sufficient manpower for the Civil War, World War I, World War II, and for more than half of the Cold War.
- "A system of selective service was mandatory for these modern, protracted conflicts."[51]
- "The use of conscription spurred volunteer enlistments, not only for the Regular Army, but also for the Reserve and Guard. During each period, the presence of a draft prompted men to volunteer."
- "With the exception of periods of the Civil War and the last 4 years of the Vietnam War, politicians and the public widely accepted the use of the draft. Protest against the draft was minimal during most of the time that the nation used this manpower policy."[52]

Yet, when people discuss the draft today, they tend to remember how the class-biased draft of the Vietnam War helped drive Americans apart and forget the more equitable draft that existed during World War II and for most of the Cold War, which helped bring the country together.

There are also two insights from the creation of the AVF. First, the "Volunteer Military Posture" to replace the draft was supposed to be "accompanied by a robust standby draft system as a hedge against a large or protracted war requiring manpower in excess of the AVF's capacity to provide." This system has never been established. The only step taken in this direction was on June 27 1980, when Congress, at President Jimmy Carter's request after the Soviet invasion of Afghanistan, enacted a law requiring all 18-year-old men to register for possible military service. This is a far cry from a scenario in which "young men would continue to register and be examined and classified, creating a yearly reserve pool available for immediate induction."[53]

Second, "the essence of a volunteer military is that it obliges no one to serve his country." While the creation of the AVF deflated political opposition to the Vietnam War, it also weakened the idea of a universal military service obligation. The focus was on a market-based approach with the right pay and benefits as the impetus prompting volunteers to enlist and serve. It has also freed politicians to use the military more as an answer to strategic questions without obliging the rest of the nation to suffer the consequences of such strategic decisions.[54] It affects only those "who wanted to volunteer for this [the military] anyway," without committing the vast majority of the U.S. public to any kind of service or sacrifice.[55]

Notes

The author is indebted to the contributions of Dr. Jonathan Shay, Dr. Bruce I. Gudmundsson, and Dr. Steven Canby over the past eight years regarding how units should be manned and sustained through recruiting and conscription.

1. Quoted in Jeffrey Record, "Equal Obligation versus Equal Opportunity: The Case for Conscription," in *The Anthropo Factor in Warfare: Conscripts, Volunteers, and Reserves* (Washington, DC: National Defense University, 1988), p. 225.

2. Donald E. Vandergriff, *Path to Victory: America's Army and the Revolution in Human Affairs* (Novato, CA: Presidio Press, May 2002), pp. 23–114.

3. Martin van Creveld, *Fighting Power: German and U.S. Army Performance 1939–1945* (New York: Praeger Paperback, 2007), pp. 87–93.

4. General Paul F. Gorman (USA, ret.), *The Secret of Future Victories* (Arlington, VA: Institute of Defense Analyses (IDA), 1992), pp. 26–53.

5. Hearings before the Subcommittee of the Appropriations, House of Representatives, 79th Congress, 1st Session, *Military Establishment Appropriations Bill for 1946*, p. 9.

6. In World War II, the Marine Corps expanded 25-fold; the Navy, 30-fold; and the Army, 60-fold.

7. Van Creveld, *Fighting Power*, pp. 65–70. Also see William S. Lind, *The Maneuver Warfare Handbook* (Boulder, CO: Westview Press, 1984).

8. The 35 years of draft represent 2 years during the Civil War (1863–1865), 1 year during World War I (1917–1918), 5 years during World War II (1940–1945), and 27 years during the Cold War (1945–1947 and 1948–1973).

9. U.S. Army, *The Personnel Replacement System in the United States Army* (Washington, DC: Department of the Army, 1954), p. 8.

10. John C. Fitzpatrick, *The Writings of George Washington from the Original Manuscript Sources, 1745–1799* (Washington, 1931), p. 163. Also see *The Personnel Replacement System*, pp. 11–12.

11. George Flynn, *Conscription and Democracy: The Draft in France, Great Britain, and the United States* (Westport, CT: Greenwood Press, 2002), p. 11.

12. Stuart H. Altman and Alan E. Fechter, "The Supply of Military Personnel in the Absence of a Draft," *The American Economic Review*, vol. 57, no. 2 (Papers and Proceedings of the Seventy-ninth Annual Meeting of the American Economic Association, May 1967), pp. 19–31.

13. Colonel William M. Raymond. "Uncle Sam Says, 'I Want You!'—The Politics of the Draft and National Service," *SAMS Monograph* (Fort Leavenworth, KS: Command and General Staff College School of Advanced Military Studies, 2005), p. 12.

14. Russell F. Weigley, *History of the United States Army* (New York: Macmillan Publishing, 1967), p. 208.

15. Allan R. Millett and Peter Maslowski, *For the Common Defense: A Military History of the United States of America*, revised and expanded (New York: The Free Press, 1984/1994), p. 207; and U.S. Department of the Army, *History of Military Mobilization in the United States*

Army, 1775–1945, by LTC Marvin A. Kreidberg and First Lieutenant Merton G. Henry, Department of the Army, Pamphlet No. 20-212 (Washington, DC: U.S. Government Printing Office, 1955), p. 104. Hereafter cited as Kreidberg.

16. Raymond, "Uncle Sam Says," p. 13.

17. Raymond, "Uncle Sam Says," p. 14.

18. Robert K. Griffith, Jr., "To Raise and Support Armies: The Evolution of US Military Manpower Systems," in *The Anthropo Factor in Warfare: Conscripts, Volunteers, and Reserves* (Washington, DC: National Defense University, 1988), p. 21.

19. Weigley, "History," p. 342.

20. Ibid., p. 343.

21. Mark Sullivan, "Conscription," in *The Military Draft: Selected Readings on Conscription,* Martin Anderson and Barbara Honegger, eds. (Stanford, CA: Hoover Institution Press, 1982), p. 29.

22. Raymond, "Uncle Sam Says," p. 18.

23. Ibid., p. 19.

24. Ibid., p. 20.

25. Sullivan, "Conscription," p. 30.

26. Raymond, "Uncle Sam Says," p. 19.

27. Raymond, "Uncle Sam Says," p. 20.

28. John G. Clifford, "Greenville Clark and the Origins of Selective Service," *Review of Politics*, vol. 35 (January 1973), p. 17.

29. Raymond, "Uncle Sam Says," p. 26

30. Flynn, *Conscription and Democracy*, p. 18.

31. Gus C. Lee and Geoffrey U. Parker, *Ending the Draft: The Story of the All Volunteer Force* (Alexandria, VA: Human Resources Research Organization, 1977), p. 13.

32. Millett and Maslowski, p. 429.

33. Flynn, *Conscription and Democracy*, p. 228.

34. Lee and Parker, p. 17.

35. Ibid., p. 16.

36. Eliot A. Cohen, *Citizens and Soldiers: The Dilemmas of Military Service* (Ithaca, NY: Cornell University Press, 1985), pp. 155–156.

37. Millett and Maslowski, p. 516.

38. Richard Gillum, "The Peacetime Draft: Voluntarism to Coercion," in *The Military Draft: Selected Readings on Conscription*, Martin Anderson and Barbara Honegger, eds. (Stanford, CA: Hoover Institution Press, 1982), p. 112.

39. Lee and Parker, p. 23.

40. *Congressional Quarterly Almanac 90th Congress, 1st Session, 1967, Volume XXIII* (Washington, DC: Congressional Quarterly Service, 1968), p. 260. The votes for 1955, 1959, and 1963 are found in each *Congressional Quarterly Almanac* for their respective years.

41. Griffith, Jr., "To Raise and Support Armies," p. 19.

42. Marc Magee, *From Selective Service to National Service,* Washington, DC: Progressive Policy Institute Policy Report (July 2003), p. 2.

43. Raymond, "Uncle Sam Says," pp. 33–34.

44. Lee and Parker, pp. 29–30.

45. Raymond, "Uncle Sam Says," pp. 33–34.

46. Flynn, "The Draft," p. 263.

47. Lee and Parker, p. 38.

48. Clifford and Spencer, pp. 227–228.

49. Raymond, "Uncle Sam Says," p. 35.

50. Robert D. Foord, "Volunteer Draft Could Replenish Military Ranks," *Harrisburg Sunday Patriot News* (Harrisburg, PA: January 9, 2005), p. F1.

51. Raymond, "Uncle Sam Says," p. 37.

52. Ibid., p. 37.

53. Record, p. 235.

54. This was one of the main causes behind the "Abrams Doctrine" named for General Creighton Abrams, hero of World War II and last commander of the Military Assistance Command Vietnam. Abrams became Army chief of staff in 1973 after General Westmoreland and determined the best way to avoid dividing the nation during war was to shift a majority of Army support and service support units to the Army National Guard and Reserves in order to force the commitment of the entire nation if the decision were made by policy makers to commit to war. If the nation went to war, then more of the nation's citizens would be committed by that decision.

55. Raymond, "Uncle Sam Says," p. 38.

CHAPTER 3

U.S. Society's Impact

> Why do our senior military leaders put out this "we can't be beaten" bilge? Because they are chosen for their willingness to tell the politicians whatever they want to hear. A larger question is, why do the American press and public buy it? The answer, I fear, is "American exceptionalism"—the belief that history's laws do not apply to America. Unfortunately, American exceptionalism follows Spanish exceptionalism, French exceptionalism, Austrian exceptionalism, German exceptionalism and Soviet exceptionalism.
>
> William S. Lind "The Reality Gap"[1]

Society is largely divorced from the impacts of a changing operating environment, which, in turn, drives dangerous complacency, leading to a refusal to implement necessary reforms beyond those imagined on PowerPoint presentations.[2]

Significant and rapid change is possible, however, in times of crisis, when there is greater doubt about the efficacy of existing systems and approaches. But incremental changes at the margins remain the rule as long as the hierarchy and existing processes are generally viewed as getting the job done in a satisfactory manner. Basically, as long as the external environment is not a threat to the society itself, or significantly impacts the freedom of the people, then change agents in this environment are likely to be marginalized or worse unless they are seen to bring other assets that are valued in the existing order. Commercial organizations attempting major changes encounter the same problem.[3]

But the nation has not suffered from what Lieutenant Colonel Isaiah Wilson defines as "shock." A significant shock is an organic part of the U.S. process of change. The same holds true for the U.S. Army. While evolution is constant, revolutions are necessarily spurred by some dramatic "shock" that launches the Army into change, and even then, it may not be enough if it only impacts the Army and not the larger society. Yet, the evolving operating environment, Fourth Generation War, brings the possibility of threats from unheard of sources. For the first time in history, there may not be an ability to react to a "shock"; the "shock" may be so damaging to society that recovery from its occurrence may be impossible.[4]

Professionals should desire a divorce of the larger society from the dangers of changing environments and their threats, but at the same, time, their very profession must advocate changes (revolutionary ideas through evolution) that allow the Army to adapt to and counter these changes. The trick is being able to explain to the political leadership why the profession of arms has to remain divorced from those factors that are taking place within society, and also lead to the pursuit of ideas that may impact the profession as well as the effectiveness of the Army. At the same time, the Army profession will continue to abide by and respect political control as stated by the U.S. Constitution.

The Leader Crisis in America

The response by many leaders—political and military figures in positions of power that can influence, create policy, and dictate strategy—in addressing almost any military problem has become a tradition: "throw money at it." In terms of security issues, this allows the military to buy more complex and advanced technology driven by the Revolution in Military Affairs (RMA), which is based on studies funded by organizations that, in turn, are paid by those in power to take data and make them say what supports the beliefs of the regime in power.

Proposals and choices by "the establishment," "inside the beltway," in Washington, DC, are accepted in silence by the citizens, whose focus is on less complex issues. Those in power see this silence as consent. As a result, the worst strategic decision in the history of the country was made when we invaded Iraq. In his book, *The New American Militarism*, Andrew J. Bacevich writes that "we can avoid future disasters by understanding how our doctrines went wrong and by returning to the precepts laid down by our Founding Fathers, men of infinitely more wisdom than those currently holding reins of power."[5]

On top of this problem, the U.S. budget deficit has grown dangerously high, and economic issues such as an impending energy crisis are not adequately addressed, while obligations made to past generations called Social Security have been raped in order to pay for the short term, threatening again to undermine promises made decades ago and supposedly upheld since then.

Despite an unprecedented access to knowledge through the Internet and television, most citizens choose not to dive into the details of knowing the reality of the world around them. It is not their fault. Our citizens mean well, but they are so busy in pursuit of the "American dream" that they have been conditioned that they only have time to deal with trivial issues—what Paris Hilton or Britney Spears is doing dominates mainstream news media, while real global issues take a back seat. Complex issues are substituted by entertaining and easier to understand subjects such as political scandals or the sexual adventures of the rich and famous. These issues dominate the airways and headlines. They take less of a citizen's limited time to understand and thus are easier to deal with. Today's elite establishment encourages an otherwise well-educated public to remain in ignorant bliss of the realities occurring around them.

Today's ignorance by choice is due to an evolution of intertwined social factors that have combined to dangerously isolate our nation from the outside realities of a world in trouble. The phenomenon may be considered the "cheerleading effect." The irony is that the cheerleading effect is an unintended tumor of our democracy combined with the cancerous and unplanned indoctrination by television. Another irony behind "cheerleading" is our country's unparalleled triumphs amid a world history of turmoil.

The United States has experienced the highest levels of prosperity seen in the history of mankind. Its citizens, through the Constitution, practice almost unlimited freedoms within the constraints of laws that are fairly enforced. Its citizens can travel across 3,000 miles of the nation from the Atlantic to the Pacific oceans at will and without fear. People of different colors and races live among each other with little violence and almost no racism. Free enterprise, capitalism, and almost unlimited resources have given us a lifestyle that is the envy of the world. But this achievement has led to a dangerous self-righteousness, particularly among U.S. leaders—the new elite—where selfless service has been replaced with selfish service— "what is in it for me?"

The duty has now become: Keep the message positive; keep the ship on course, no matter where the course may lead, as long as the people believe it is only going to get better at minimum cost to both the leadership and the public.

> American attitudes have changed since the days of "Be All That You Can Be," when modern recruiting saw its heyday. Some say we're living in the age of Club America. The job of club management is to keep problems away from the members who pay dues to pursue a life of their choosing. Military service doesn't fit into this equation. Simply put, military service isn't "cool" by today's standards, and parents seem to go along with this assessment. The nation's political leadership—starting with the President and including members of Congress—must make it cool again.[6]

When politicians do stray from the course and tell it like it is, they are punished in the polls, as was the case when presidential contender Senator John McCain (R-AR) lost in Michigan. Senator McCain told the people that their automobile jobs would not return, that they would have to adapt to changing times, and then he proposed a plan of re-education to prepare them for these changes. Senator McCain's rival, Governor Mitt Romney, on the other hand, told the people what they wanted to hear: that the automobile industry was only in dire straights as a result of government meddling and that he (Romney) would return them to the good old days, if only they would elect him as president.

> Romney may not have won in Michigan so much as McCain lost it. And he lost it because of a characteristic tendency that makes him Romney's

> opposite—political rigidity based on a sense of his own personal rectitude. Having said jobs in Michigan were not coming back, he went to Michigan and praised efforts to mandate an increase in fuel-mileage standards, which auto executives claim will raise the price of a car fully $6,000—a job killer, in other words. And he spoke against drilling in the Alaska National Wildlife Refuge, which is the only realistic way for the United States to increase its own domestic oil supply.
>
> McCain's line is that he is a straight talker. But there are moments he seems to make a fetish of his own honesty, and asks others to support him solely because of it.[7]

Publicly admitting that there are major problems that demand action and possible public sacrifice are also likely the same reason not one presidential contender, except for Congressman Ron Paul, has addressed how he or she is going to fix Social Security or Medicare, for which the costs are spiraling out of control. Telling the truth will only get politicians in trouble and not elected—the people will punish them for telling the "inconvenient truth."[8]

Over time, nations or large organizations such as an Army take one of two evolutionary routes: continue to adapt with changing environmental conditions in order to sustain their core beliefs or become complacent, resting on one's laurels. If it is the latter, it is because its very success has led to its isolation from a changing environment. Decisions based on facts and assumptions that led to success become dated. In turn, the organization or nation should demand that its leaders conduct critical analysis in order to propose courses of action to meet goals.[9]

The right course of action may force leaders to make hard choices that, in turn, require they ask those they lead to change their accepted habits and even make sacrifices. Rather than accepting and attempting to adapt the current and positive way of life to the unknown problems the future may bring, those in power and the citizens at large may choose to react to the demands of the environment by doing nothing; keeping things as they are, while pretending to do something through colorful rhetoric and complicated presentations. Then, in order to justify doing nothing or not making hard choices, a culture adopts the "cheerleading effect."

The cheerleading effect is where the only accepted message is a positive one. Everyone in a culture that accepts cheerleading wants to feel great and wants the party to go on, even after everyone has forgotten what the celebration was for. It becomes an endorsement of current accepted beliefs or policies, and over time, it even transforms history into one-sided views that become traditions and myths that, in turn, support the current accepted message.

Cheerleaders are motivators and entertainers. Cheerleaders are also attractive. At football games their primary duty is to try to get the crowd involved in the game, particularly in home games where crowd noise can be used as the "12th player." Hopefully, this motivates their team, and if crowd noise is loud enough, it makes it hard for the visiting team to call or change plays and hear the snap count.

Cheerleaders also provide entertainment. They are athletically fit and do numerous acrobatic stunts to keep people motivated during down times. Even in the direst moments of a losing contest, cheerleaders try to convince people that through some miracle, their team is going to win. Only after the loss in a big game do you ever see cheerleaders show signs of despair.

When cheerleading is used by an organization to undermine and block change, then it is not used to motivate, but to lie and eventually destroy members' trust in the organization.

"Cheerleading effect" becomes a cultural norm as it accompanies and justifies complacency. It is particularly evident in democracies in which the nation periodically elects new leaders or endorses current leaders by allowing them another term in office. In turn, the politicians promise to do several things to take care of their followers. To the individual, cheerleading provides security, even though, over time, it eventually may lead to the demise of the culture. Another aspect of U.S. culture that supports the cheerleading effect is our focus on the individual.

Professions also have cheerleaders. They are the members who keep the profession upbeat and retain the faith in the doctrine of the organization. This is especially needed in bad times, or when senior members of the profession try to sell ideas that may be seen as unpopular with the rank and file. Most of the time, those who are cheerleading are also the ones that have benefited the most by how the profession works, especially in terms of how to be successful. Thus, slowly over time, cheerleading receives a place in the uppermost level of the profession. Today, cheerleading is enhanced with the use of video technology. While cheerleading's primary use is entertainment at athletic events, it becomes a dangerous phenomenon when it forces members to contradict the proven values of the profession.

Complacency in an Army is very dangerous, especially if successes drive an Army, and its larger society, into a mindset that moves beyond mere compliance. There is a chance that the cheerleading used to drum up support for a war or mission becomes the truth if that particular event is successful. At this point, everyone, but particularly decision makers, starts to believe that a "perfect" model has been found and is applicable for any task. Any debate or argument is seen as disloyalty.

Cultures even evolve a system of incentives that only rewards behavior that is seen as favorable in maintaining the status quo. For whatever reasons, everyone gets caught up in doing whatever it takes to keep the victory celebration going. Eventually, "committing the truth" is not accepted if it contradicts positive themes. Shakespeare shows an extreme example of what happens to a culture in the play *Coriolanus*, when the central character is not allowed to point out problems because the people are only used to good news.[10]

Coriolanus tries to argue with other Roman senators regarding telling the truth about major issues they must fix. In turn, the senators try to convince him not to tell the Roman people the truth. Coriolanus replies by pointing out the dangers in the contradiction between what he has been taught to do as a leader, which is to identify problems and then recommend a solution, and what they are asking. He fails to convince the senators despite his pleas regarding the dangers of doing what has become an accepted norm—lying.

> No, take more: What may be sworn by, both divine and human, Seal what I end withal! This double worship, Where one part does disdain with cause, the other Insult without all reason, where gentry, title, wisdom, Cannot conclude but by the yea and no Of general ignorance,–it must omit Real necessities, and give way the while To unstable slightness: purpose so barr'd it follows Nothing is done to purpose. Therefore, beseech you,– You that will be less fearful than discreet, That love the fundamental part of state More than you doubt the change on't, that prefer A noble life before a long, and wish To jump a body with a dangerous physic That's sure of death without it, at once pluck out The multitudinous tongue; let them not lick The sweet which is their poison: your dishonour Mangles true judgment and bereaves the state Of that integrity which should become't, Not having the power to do the good it would, For the in which doth control't.[11]

Shakespeare uses the play *Coriolanus* to demonstrate, to an extreme, the choices those who "commit truth" have when their display of moral courage is confronted and punished. As most members of the profession unknowingly accept the "easy wrong over the hard right," they find comfort in cheerleading. They do not realize that their security may only be short-lived while the going is good. Instead, the majority move to protect the organization's cheerleading without question, believing that nothing will ever change. They begin to do what it takes to protect their sanctuary; they label and drive away those who "commit the truth."

Labeled by terms such as malcontent, rabble rouser, or even extreme nouns such as zealot, truth tellers have two choices. They can comply and cave in morally, or quit or resign from the profession. In order to maintain their pride and uphold their character, they can still fight the organization from the outside (where they normally have less impact) or go and do something else all together. Over the long term, driving out those "who commit the truth" drives professions to look more and more inward versus outward when adapting to the changes in the environment in which they exist. Cheerleading would not, by itself, be a dangerous phenomenon when the culture maintains a values system, but when tied to society's new definition of success, it begins to assist in the deterioration of the state.

The geographic isolation of the United States from most threats allowed for the nation's undisturbed prosperity. The assistance of a large entertainment industry allows people to be constantly entertained and escape from constant pressure to achieve status through wealth. The impact has changed the culture's definition of success from being a good citizen who raises one's family, pays taxes, and abides by the nation's laws to how much money people earn.

Success is now measured by how much wealth one has amassed, displayed through material goods. The pressure is there every day to achieve this success, competing with everyone. It explains the myriad of corporate scandals that occurred in the early years of the twenty-first century, record-high consumer debt, and President Bush's call for citizens to spend more in response to the attacks on the United States on September 11, 2001. Combined with the impacts of cheerleading on the culture's leaders, this trend into self-serving leadership has impacted how leaders are chosen and promoted. The game then becomes about retaining one's position, status, and power, which, more importantly, are all tied to earning power.

Cheerleading, combined with the new definition of success, infects an Army with even more deleterious effects because it must be able to adapt and change rapidly with the changing face of war. Only leaders of character will solve today's and future problems, and the Army must lead the way in developing the core of our nation's future leaders from what exists today.

The Future Soldier?

Baby Boomer officers thought Generation X caused enough ruckuses when they entered the Army; their 40 million memberships are paltry compared with "Gen Y's" 68 million. Because of their numbers, education, and abilities, "Gen Y" will influence the Army for years to come.

Generation Y is the most educated generation ever but, as is common with potential officers, unless they have prior service, they have limited practical experience. They are, however, willing to learn. In studies that can give insight to our future cadets, Randstad's *Employee Review* found 6 out of 10 Generation Y say their workplace's on-site/internal training is important in their decision to stay, yet only 43 percent say they're satisfied with the opportunities they have to learn new things in their current job. That compares with 55 percent of Generation X, 53 percent of Baby Boomers, and 56 percent of mature workers who say they're satisfied with learning opportunities at work. What is the implication for the Army?[12]

Generation Y's dissatisfaction with training opportunities spills over into their overall satisfaction with their jobs. Only 43 percent of Generation Y workers say they get a lot of personal satisfaction from the work they do, compared with 65 percent of Gen Xers, 68 percent of Boomers, and 77 percent of Matures. "Generation-Y is the future of the Army," Jim Reese, CEO of Randstad North America said. "The Army needs to know how to attract them and challenge them." He said that "allowing Generation-Y people to participate in decision making, to challenge them, train them and allow them to 'to play hard as well' from time to time will enable the Army to reap big dividends from America's potential leadership." This generation offers many challenges to recruiters and initial entry trainers.[13]

One of the refrains heard from small unit leaders is that new soldiers lack skills, particularly life skills. They tell us that many young people graduate from high school or college without knowing the basics: literacy, numeracy, and what it takes to complete a mission or balance a checkbook. This holds true with both enlisted and officers. In a study by the Pew Charitable Trust, 50 percent of college graduates were found to be lacking in simple life skills. These grads were tested for three types of literacy skills: analyzing news stories and other prose, understanding documents, and having math skills needed for balancing checkbooks or figuring restaurant tips.[14] A new study just released by the American Institutes for Research (AIR), a behavioral and social science research organization, echoes Pew's findings. More than half of the students nearing graduation at four-year colleges lacked the literacy skills to handle complex tasks, such as analyzing arguments in newspaper editorials or understanding credit card offers.[15]

Students nearing graduation at two-year colleges scored even worse, with 75 percent not having the basic skills. This AIR study was the first to target students closing in on graduation. The researchers utilized the same test as the National Assessment of Adult Literacy, the U.S. government's examination of English literacy among adults. Results of that government study, released in December 2007, were similarly disappointing. The results reflected about 5 percent of U.S. adults are not literate.[16]

All of these studies raise serious questions about college curricula as well as the skills students possess when they arrive at college. The Army is finding that it has to test all of its officer candidates for basic literacy and numeracy. Sixty-eight percent of applicants failed this basic assessment. All of these outcomes do not bode well for the United States' ability to engage in Fourth Generation Warfare.

In summary, members of Gen Y—the potential soldiers of the future—tend to be optimistic and values oriented, and have a desire for achievement. They are educated, familiar with emerging technology, and computer literate. They are well behaved and work well in a team environment. They believe in themselves, invest in themselves through education, and tend to be benefits oriented. They look for a balanced life and, although they feel favorably toward the military, they can be reluctant to make long-term commitments. Gen Y is less physically fit and more obese than previous generations. Like the generations that preceded them, if they come into the Army, they will come in to succeed, not fail.

Competition Is Sharp

The Army is in competition worldwide, and employees worldwide are stressed. The Army is seeking the same skilled workers that companies need to manufacture their products and serve their customers. Some companies have had mission-critical jobs open for more than a year. One hospital has been looking for a physician for three years.

The demand for innovation is not going away, particularly in Fourth Generation Warfare, where innovation is a must. This fuels an ever-increasing demand for soldiers with the right skills to make innovation pay off. Because the United States and global economies are competing for the same skill sets and abilities, long-term, the job market is expected to be strong, placing more demands on the Army and its recruiters, but also leaders to retain these soldiers. The fact is, when the demand for services accelerates, the Army is going to be facing a sustained soldier shortage, due in large part to a seismic demographic shift on the horizon.[17]

Some 76 million Baby Boomers–the largest generation in the workforce today – are gearing down for retirement. Their departure is expected to drive down the unemployment rate to a new 30-year low of 3.9 percent ... or lower. The Department of Labor projects that by 2010, there will be 10 million more jobs than there will be workers to fill them. And that's factoring in the nearly 7 in 10 "Mature" workers who, according to studies by the AARP and *USA Today*, CNN, and Gallup, plan to stay in the workforce and work at least part-time past traditional retirement age.[18]

By 2030, the American workplace may be 35 million short of the 57 million new employees it will need to maintain current annual growth in per capita gross domestic product. Many who remain in the workforce will lack the basic, as well as specialized skills, to perform digital-era work. Seventy percent of the fastest growing jobs will require more than a high school education, and 60 percent of jobs will require skills that only 20 percent of workers currently have. So not only will there be a labor shortage, but there will be a skills shortage as well.[19]

The Army will be in an *all-out war for talent*.

What's behind this storm is a combination of years of low birthrates in the developed countries with unprecedented global prosperity and shortages of trained workers. Complicating this difficult situation is the competency deficit and a lack of leadership, which makes recruiting for officers tougher than ever before. When the schools are not adequately preparing young people for work or college and employers haven't been training their next generation of workers, it is a setup for disaster.

Advancing technologies address some of these deficits, but they only increase the demand for more highly skilled workers, who already are in short supply. Also, technology does not address the challenges presented by Fourth Generation Warfare, the need for multi-skilled soldiers and leaders who have to know when to pull the trigger, forge an alliance with a local chieftain, or supervise building of a school, as well as possess the cognitive ability and ethical reasoning to transition to having to kill an enemy if the situation demands.

Another Factor, Root of the Culture

The approach the Army takes to man its legions—how it recruits, trains, and provides replacements—began its cultural evolution toward linearity and the Newtonian way of war alongside the emergence of the large corporations and industrial captains of the late 1800s. They embraced the theories and practices of Frederick Taylor (1856–1915), the great managerial and labor industrial theorist of the late 1800s and early 1900s. Taylor was explicit about his intentions to use "scientific management" to control capitalist means and conditions of production. This theory and approach were referred to as Taylorism.[20]

Later, these methods evolved into the Lanchesterian models, which were scientific tools of trend analysis, determination, and sample means used to predict the outcome of battles. In the Army, this is referred to as "the arithmetic of the battlefield" or "battlefield math." The "Lanchester" equations modeled the outcome of battles as a function of force ratios, focusing on losses and the attrition of forces only. In turn, this told the Army personnel system what type and how many "parts" it would need.[21]

The secretary of war from 1899 to 1904, former lawyer Elihu Root, embraced Taylorism, evolved using the recommendations of U.S. Army Maj. Gen. Emory Upton (1839–1881). Upton was the leading reformer in trying to move the Army from the frontier to full-time professionalism based on the Prussian model. Nonetheless, Root's interpretation of Upton's work (he died before Root took office) followed the preaching of Taylor.[22] As a result, Root's reforms moved the U.S. Army from a constabulary force, reliant on an undependable militia system, to a force that could fight in a major war with and against the armies of European nation-states.

The Army followed the corporate world example, structuring its activities in a framework based on Frederick Taylor's theories of managing and training workers, as discussed earlier. Taylor's theories were influenced by Newtonian determinism, which applies linear thinking to the development of leaders.[23] Taylor's greatest contribution to production efficiency was to break down complex production tasks into a sequence of simple, standardized steps. This permitted him to design a standardized mass production line around a management system that classified work into standard tasks and workers into standard specialties. This combination established work standards and the people who were trained to these standards became interchangeable cogs in the machine. This greatly simplified personnel management in a vast industrial enterprise.

To be sure, Taylorism transformed industrial production, but it also had a dark side: Taylorism treated people as unthinking cogs in a machine. By necessity, these people had to accept a social system based on a coercive pattern of dominance and subordinance and centralized control from the top. Every action and every decision made in the organization was spelled out in the name of efficiency. In theory, the entire regimen flowed from the brain of one individual at the top of the hierarchy.

A complementary management dogma also emerged during the Progressive Era. This was the theory of "ethical egoism," which asserted that all people are motivated solely by self-interest. By extension, all people would respond predictably to a variety of positive incentives (money, pleasure, advancement, distinction, power, luxurious prestige goods, and amenities) or negative incentives (which took the primary form of a fear of losing the positive benefits, but also outright punishment and pain).

The two previous modern phenomena were laid on top of the final factor that had already existed during the Industrial Age as the education philosophy of René Descartes. Descartes was a famous mathematician who broke down engineering problems in sequence, making it easier to teach formulas to engineering students. This approach was translated into French military training, where the French found it easy to break down military problem solving into processes (checklists) to educate their officers and their awaiting masses of citizen soldiers upon mobilization. This approach, in turn, was copied by the U.S. Army as a way to turn mobilized citizens into soldiers and leaders quickly.[24]

The Cartesian approach allowed the French (and later the United States) to easily teach a common, fundamental doctrinal language to many who were new to the military. It significantly reduced the time it took to master basic military skills. The downfall of this approach is that it simplifies war (complex problems) into processes in which the enemy is only a template, not a free-thinking adversary.

The Cartesian approach also slows down a decision cycle by turning the planners' focus inward, on process, instead of outward on the enemy. The problem with this approach is that it's hard to adapt to whatever problems are at hand. It is the same thing with operations research, which is a powerful tool, but only for solving certain well-defined problems. The problem with the U.S. military is that it applies a mechanical approach to all sorts of inappropriate problems, including leader development, specifically education. The French, and then the United States, while relying on a massed citizen army in the late nineteenth and early twentieth centuries, had to find a way to instruct many citizen officers quickly in military doctrine.

Additionally, because of the casualties of World War I and the advance of modern weaponry and its destructiveness, the French needed a way to teach its officers how to control these resources to concentrate firepower so they could compensate for their lack of unit skills on the battlefields. They used an orderly and systematic approach to planning that was similar to the Military Decision-Making Process.[25]

When the new U.S. Army arrived in Europe in 1917, it was led largely by citizens who had been transformed into officers almost overnight, and its soldiers needed to learn the fundamentals of the profession of arms quickly. All U.S. staff officers and commanders attended French schools in planning and controlling forces in combat. The United States and France were the victors in World War I and saw that victory as a validation of their training process. When the French developed the doctrine of "Methodical Battle" in the interwar years, the United States copied it with all its accompanying process-focused education. The U.S. Army carried this over to its education and training, as well as its doctrine.[26]

France and the United States practiced Progressive Era personnel theories and opened the net wide to accessions in order to be democratic and fair. Missing from this practice was a hard "filter" to judge character under stress prior to awarding a commission. It was felt that new officers could learn on the job either in peacetime duty or in war. This was a very harsh way to develop and prepare leaders. Both countries felt that they were the victors in war without examining why they were the victors, and they ignored mistakes and lessons learned.

In sum, the combination of policies and beliefs that Descartes, Taylor, and Root began in the Progressive Era are seen today through easier accessions, faster promotions, no obligation to attend essential leader training school, the refusal to allow cadre to kick out nonperformers, and quicker pay raises. These policies are fully consistent with their theory of human behavior and they stem from the idea that a system has a need to fill spaces with faces.

Taken together, the ideas that people are interchangeable cogs in a machine and that self-interest is the only significant motivator of behavior. This helps explain why the Army thinks that increasing its "production" of lieutenants, cutting out necessary training for young leaders, and reducing the promotion time to the rank of major will solve its statistical readiness issues with deploying units. The Army focuses on meeting near-term requirements mandated by the Army and Congress for field grades rather than solving potential retention problems in the long-term.

The ideas of Descartes, Taylor, and Root dominated management science and War Department circles a century ago, but their ghosts are haunting the Army's Human Resources Command and the Deputy Chief of Staff for Personnel today. Moreover, the ghosts of Descartes, Taylor, and Root will continue to haunt the Army's personnel manager, and filter down to those tasked to develop leaders, as long as Congress shows no interest in rooting out causes of the Army's personnel crisis.[27]

But Congress and the press are blinded by the sterile promises of another techno-centric analogy mentioned in the previous chapters—the RMA—that is based on the idea that war is a mechanistic process and that machines are the true source of military prowess (as if U.S. opponents stand in the open and let us kill them all day). It was the RMA Army that we took to war with Iraq. The RMA is the specter of Descartes, Root, and Taylor that haunts the Army today.

There are dangers of reasoning by analogy. Used properly, analogies are powerful reasoning devices because they unleash the genius of imagination and creativity, Einstein's thought experiments being cases in point. But analogies are also very dangerous; on the one hand, they can simplify complex problems and capture our imaginations, but on the other hand, when used improperly, they shackle the mind and take it off the cliff.

Believing that the Army is like a business or that good business practices will solve military problems are examples of misplaced analogies that take its leaders off the cliff. Effective business practices are often very different from effective military practices. This is particularly true in the area of personnel policies, where the idea of soldierly virtue embodies the ethos of self-sacrifice, and where, as Napoleon said, the moral is to the material as three is to one.

Descartes' and Taylor's management theories were linear, and they impacted adaptability and innovation in leaders by helping to solve personnel shortages, which in turn impacted quality of leadership. Standards in officer accessions (how individuals are prepared to become officers), leader development, promotions, and attendance to military and civilian education opportunities, were lowered to meet the need for "bodies" or "spare parts." There is a mismatch between input and output using this system; namely, junior officers are now "punching out"—leaving—faster than the equivalent number of lieutenants promoted to captain or major based on the predictions by personnel operations research systems analysts for the future need for majors.

Why would the combat-savvy junior officers want to punch out after doing what they had been prepared to do? The most frequent complaint made by the best and brightest of the numerous lieutenants and captains about departing is that they are frustrated with being micromanaged to death by those immediately above them "trying to make their mark" in a top-down, hierarchal, PowerPoint-driven culture in which junior officer potential for increased responsibility is limited by too many officers "waiting their turn." It is a culture that does not give them enough time in leadership and command positions to do the things they joined the Army to do: for instance, learning the art of soldiering, such as troop command and tactical leadership. Even more so, they are fed up with and insulted by the lowering standards that are appearing across the ranks, from which the Army recruits, commissions, and promotes.[28]

The Army solution: balance input with output by pumping up the input—in this case by beginning to demand more from accession sources. This means that while Recruiting and Cadet Command raises its "mission," but with fewer resources, the future battlefield requires a higher quality leader and soldiers. Other solutions imposed by the Army include raising the percentage of leaders who are promoted to major (now 99 percent) in an effort to speed up accession rates and, in 2004–2008, lowering recruiting standards that include more waivers for unethical and immoral behavior, such as accepting recruits with criminal records, or passing on lieutenants who are caught cheating on examinations in an Army course.

A linear approach does two things that are undermining the long-term health of the Army and its ability to fix today's problems in an era of Fourth Generation Warfare, when the Army is asking lieutenants and sergeants to make decisions with strategic implications. First, this becomes a culture that tries to bring control to uncertainty and develops a technocratic culture. Second, a culture that trains and manages for predictability and "one right answer" subdues a leader development system that favors critical thinking among chaos.

The Army's solution is akin to increasing the size of the bilge pump rather than plugging the hole that is sinking the ship. This kind of bilge also helps us to understand why the first four letters of the word "analyst" are "anal." Why is this happening in the twenty-first century? The Army has viewed the management of its people through the tired old eyes of Secretary of War Elihu Root and the turn-of-the-century industrial theorist, Frederick Taylor. It has become as acceptable as fish is to water; it's just the way it is done.

The Army's practice to retain soldiers and officers by promoting them faster aims to solve a structural problem by bribing people to stay—the positive incentive of faster promotions will buy their loyalty, patriotism, and the moral strength to go in harm's way. Yet this kind of appeal to self-interest is precisely the kind of policy that has failed repeatedly in the past and will actually increase the exodus of the Army's "best and brightest" young people—thus robbing the Army's future. It is based on the dehumanizing assumption that its officers (and noncommissioned officers) are mindless, undifferentiated, replaceable cogs in a machine.

Finally, these reforms also took into account the primacy of the individual that remains at the heart of the U.S. Constitution. Root's new system could not appear to threaten American democracy by postulating a military that was separate and isolated from civilian society. Root combined the popular corporate management practices of Taylor with the emerging theories of personnel management in the Progressive Era. Root's system promised individualism because it appeared to give everyone a "level playing field" in which to move up to the top.

For the Army, Root's reforms provided the beginning of a culture that could support the distinctive mobilization doctrine of the United States. It was a doctrine that worked well as long as the United States was fighting a linear war.[29] However, the U.S. Army today, and for the near future, is not fighting linear wars. The critical aspect of Root's reforms involved personnel management initiatives that rode the wave of scientific management then sweeping the nation and the Western world. At the time Root first implemented these measures, *any* adjustments brought improvements to the bedraggled post–Civil War Army. However, what began as reform soon became a bureaucratic straitjacket that would have long-lasting, negative impacts on later generations of officers, enlisted soldiers, and the Army as an institution.

Ironically, policies that favored the primacy of the individual alongside institutional rhetoric of selfless service created a culture of conflict that few, if any, could escape.[30] These laws and supporting policies have become legacies, or "untouchables," that the Army culture sees as necessary to its very function. The legacies support an increasing motivation of individuals fueled by self-interest that pits the espoused selflessness of Army values, especially among leaders, against careerism.

Conclusion: Taylor Out, Deming In

The Army has to fire Taylor, DeCartes, and Root. It has stated it needs to, so now let's do it. Forget the fact that the Department of the Army (DA) is designed to fight America's wars. Instead, think of the DA as the world's largest corporation—a corporation that has a crisis in human resources.

If the DA were a publicly traded company, its budget would be 50 percent larger than that of retailing giant Wal-Mart and twice the size of General Motors. DA employs 740,000 people (soldiers and civilians), is spread over 146 countries, and operates 70,000 motor vehicles and 6,000 aircraft.

And yet, DA is stuck in the past. DA has a crisis in personnel management because it is still stuck in the old Industrial Age model created by people like Frederick Taylor and Henry Ford. In that model, workers were simply cogs in the wheel, easily trained, easily replaceable. Workers were not expected to think. In fact, thinking was discouraged. They were simply told when to show up and what to do. They had no say in production or engineering.

But that model has been supplanted. W. Edwards Deming's model replaced the model put forward by Frederick Taylor.[31] Deming insisted that the lowest-level workers must be empowered. In a plant run according to Deming, the lowest-level workers can and are required stop the production line, at any time, if they detect a problem that they cannot solve before the article moves to the next station. Today's biggest and most successful companies, like Toyota and Honda, use organizational models that raise the decision level of the lowest-level personnel. Deming's model of personnel management must be adopted by the Army, as we will see in the next chapters. Failure to do so will result in a failure of the Army as a whole.

Notes

I am indebted to the following people for their input and ideas to this chapter: Robert Bryce, LTC (ret.) Allen Gill, William S. Lind, Dr. Charlie Moskos, John Tillson, and Mark Lewis.

1. William S. Lind, "The Reality Gap," retrieved from http://www.d-n-i.net/lind/lind_5_11_05.htm; accessed May 18, 2005.
2. I am indebted to the online discussions where I have gathered these thoughts from Colonel Chris Paparone, Colonel George Reed, Mr. John Tillson, and Major Ike Wilson.
3. Deborah Stone, *Policy Paradox: The Art of Political Policy Making*, 2nd edition (New York: W.W. Norton, 1997).
4. Graham Allison and Philip Zelikow, *Essence of Decision: Explaining the Cuban Missile Crisis*, 2nd edition (New York: Longman, 1999).
5. Paul Craig Roberts, "The New American Militarism," retrieved from http://www.antiwar.com/roberts/?articleid=4445; accessed January 18, 2005.
6. Charles Krohn, "Atten-Shun," *Washington Times* (June 4, 2005).
7. http://flapsblog.com/?p=6303.
8. I must thank my close friend and former boss at Georgetown Army ROTC, LTC Allen Gill (USA, ret.), for these thoughts.
9. Karl E. Weick and Kathleen M. Sutcliffe, *Managing the Unexpected* (San Francisco, CA: Jossey-Bass, 2001), pp. 3–8.
10. I want to again thank LTC Allen Gill (USA, ret.), who was my last boss at Georgetown Army ROTC, and who also taught English literature at the United States Military Academy, for his insights to the lessons from Shakespeare and particularly those offered in *Coriolanus*.
11. William Shakespeare, *Shakespeare, Select Plays, Coriolanus*, William Wright, ed. (Oxford: Oxford University Press, 1891), p. 141.

12. Claire Raines, "Managing Millennials," 2002, retrieved from http://www.generations at work.com/articles/millennials.htm; accessed December 16, 2003.

13. Office of the Under Secretary of Defense for Personnel and Readiness, *Military Personnel Human Resources Strategic Plan, Change 1* (Washington, DC: Office of the Under Secretary of Defense for Personnel and Readiness, August 2002), p. 2.

14. Pew Research Center, *A Portrait of Generation Next* (January 2007), retrieved from http://people-press.org/reports/display.php3/?ReportID=300.

15. The American Institutes for Research and SRI International, *High Time for High School Reform: Early Findings from the Evaluation of the National School District and Network Grants Program* (2003), p. 4. This report, like the other reports on this page, is one of a series that will be produced as part of the ongoing evaluation of the Bill and Melinda Gates Foundation's National School District and Network Grants Program. The program provides funding and support for the establishment of new small high schools and the conversion of large high schools into smaller learning communities. This report, reflecting results from the program's second year, focuses on the design and early implementation of new and converting schools, examining what it takes to bring small school visions into practice.

16. *High Time for High School Reform*, p. 4.

17. Charles Goldman, et al., *Allocating Scholarships for Army ROTC* (Washington, DC: Rand, 1999).

18. Department of Labor, retrieved from http://www.bls.gov/cps/home.htm.

19. Alexander W. Astin, et al., *The American Freshman: Thirty Year Trends* (Los Angeles: University of California, Cooperative Institutional Research Program, 1997).

20. Jeff Minson, "Strategies for Socialists? Foucault's Conception of Power," in *Towards a Critique of Foucault,* ed. Mike Gane (London: Routledge & Kegan, 1996), p. 122.

21. Thomas Czerwinski, *Coping with the Bounds: Speculations on Non-linearity in Military Affairs* (Washington, DC: National Defense University Press, 1998), retrieved from http://www.dodccrp.org/coptin.htm.

22. Edward Coffman, *The Old Army: A Portrait of the American Army in Peacetime, 1784–1898* (New York: Oxford University Press, 1986), pp. 61–64 and 194–198. Also see John Elting, *American Army Life* (New York: Scribner's, 1982), pp. 55, 84. For information on a period of social reform between the Civil War and the Spanish-American War that preceded Root, see Jack D. Foner, *The United States Soldier between Two Wars: Army Life and Reforms, 1865–1898* (New York: Humanities Press, 1970). Foner discerns two distinct periods of reform, the first in the early 1880s and a second in the late 1880s and early 1890s. Spurred by the civil service reform movement of the early 1880s and later by what became the Progressive movement, the Army secured numerous social reforms, all aimed at making service life more attractive for the enlisted soldier.

23. John Briggs and David F. Peat, *Turbulent Mirror* (New York: Harper and Row, 1989). Also, Martin van Creveld, *Transformation of War* (Cambridge, MA: Harvard University Press, 1985).

24. S. P. Hays, "Introduction," in *Building the Organizational Society*, J. Israel, ed. (New York: Free Press, 1971). Also see J. H. Hays, *The Evolution of Military Officer Personnel Management Policies: A Preliminary Study with Parallels from Industry* (Santa Monica, CA: Rand, 1978).

25. R. Doughty, *Seeds of Disaster: The Development of French Doctrine, 1919–1939* (New York: Archon Books, 1986).

26. Major E. Swift, *Field Orders, Messages and Reports* (Washington, DC: U.S. Government Printing Office, Document UB283.A45, 1906). S. R. Stewart, *Leader Development Training Assessment of U.S. Army TRADOC Brigade Commanders* (Washington, DC: U.S. ARI Research Report 1454).

27. Donald E. Vandergriff, "Raising the Bar: Creating Adaptive Leaders to Deal with the Changing Face of War," unpublished study (Washington, DC: Georgetown University Army ROTC, June 2005). While conducting the study upon which this monograph is based, I learned that the ROTC POI (Program of Instruction) places leadership training fourth in priority behind Cadet Mandatory Training, First Aid, and Operations and Tactics for pre-commissioning tasks to be taught to cadets. Cadet Command places greater emphasis on teaching the cadets how to become soldiers than how to become leaders of soldiers. The original study recommended that Cadet Command integrate leadership tasks and training to maximize instructor contact hours. The fact that ROTC institutions have limited opportunities to train supports the review of common skills tasks taught in the formal portion of ROTC training. Another finding of this study was that the current ROTC POI does not teach the leadership attributes currently listed in Field Manual 22-100. Of the 11 leadership pre-commissioning tasks taught, none of them informs the cadets about the leadership attributes listed in current doctrine. This is primarily due to the ROTC POI being published prior to the publishing date on FM 22-100.

28. E-mail communication with several junior officers—alumni from Georgetown Army ROTC, as well as many other officers, including those who were commissioned from West Point as well as Officer Candidate School.

29. Discussions with Dr. Faris Kirkland, April 12 and May 8, 1998. Also see Faris Kirkland, "The Gap between Leadership Policy and Practice: A Historical Perspective," *Parameters* (Carlisle, PA: U.S. Army War College, September 1990), pp. 54–55; Coffman, pp. 23–40, 101–103, and 176–180; Elting, pp. 38–45, 76–78; and Russell F. Weigly, "American Strategy from Its Beginnings through the First World War," in *Makers of Modern Strategy from Machiavelli to the Nuclear Age,* Peter Parat, ed. (Princeton, NJ: Princeton University Press, 1986), pp. 439–441. The Army reformers under Upton failed to reform West Point on the German model. In 1881, General John Schofield resigned as superintendent, citing political meddling when he tried to change the curriculum from an engineering focus to a military art focus. Gradually, some technical courses were replaced with more broad, liberal, and military art as in the German system. Several officers including William T. Sherman did see the failure of the Army's organization and policies during the Civil War. A small group of officers, called "Uptopians" after General Emory Upton, attempted some type of reforms for three decades after the Civil War, and met largely with failure.

30. Jack C. Lane, *Armed Progressive* (San Rafael, CA: Presidio Press, 1978), p. 150. Also see Emory Upton, *The Armies of Asia and Europe* (Washington, DC: U.S. Government Printing Office, 1878), p. 219. Upton considered the U.S. Army experience as misguided and disastrous. He considered the overall performance of the U.S. Army in all previous wars as unprofessional and wasteful of lives. These results he attributed to ineffective policy regarding an amateur officer corps, a reliance on a militia system and faulty organization. It was General-in-Chief William T. Sherman who was behind sending Upton to Europe, specifically to study the Prussian Army, instead of sending the corrupt Belknap to Europe. John Dickinson, *The Building of an Army* (New York: The Century Co., 1922), pp. 245–270. At the time, the officer corps was not professional. Examinations and efficiency reports were used, but nothing was done when officers failed them or received a bad report. Promotions, commissions, and appointments more often were based upon favoritism and

political pull than merit. The most common argument in favor of the "up-or-out" promotion system is the experience during the Civil War and World War II, when hundreds of officers too old to perform in the field had to be replaced by younger and more vigorous officers. This argument ignores the fact that, unlike today, no mental and physical evaluations existed. Paul Y. Hammond, *Organizing for Defense: The American Military Establishment in the Twentieth Century* (Princeton, NJ: Princeton University Press, 1961), p. 10. Hammond asserts that President William McKinley brought in Root "to clean up the mess" left by the Spanish-American War. Graham A. Cosmas, *An Army for Empire: The United States Army in the Spanish-American War* (Columbia, MO: University of Missouri Press, 1971), p. 311. Cosmas claims that "McKinley strove to organize the Army on the principles elaborated by Emory Upton." Chester I. Bernard, *The Functions of the Executive* (Cambridge, MA: Harvard University Press, 1938), p. 148.

31. Kenneth T. Delavigne and J. Daniel Robertson, *Deming's Profound Changes* (Trenton, NJ: Prentice Hall, 1994), pp. 1–3.

CHAPTER 4

Achieving Parallel Evolution

"Increased centralization generates friction between the levels of command. Without well-established mutual trust, the temptation to micromanage overtakes commanders and slows decision cycles at all levels. Increased legislation works to create a rigid, mechanistic structure. All of these factors combine to create a rigid, inefficient, and sluggish organization that is slow to adapt and ultimately combat ineffective!!"

Colonel John Boyd[1]

The challenge for Army senior leaders is to build an agile, perceptive Army to deal with both existing and emerging threats. One way to do that is to minimize bureaucracy. Another is to keep the Army exceedingly close to the soldier and junior leaders in every possible manner. The third is to listen to voices on the front line in shaping Army strategy; after all, they are usually the first to see emerging threats and shifts in the evolution of war. The fourth is to protect the mavericks—the iconoclasts, those who often drive you crazy with out-of-the box ideas. The fifth is to promote people who support reinvention and innovation and back them up with rewards.[2]

In addition, leaders must discipline themselves to continually re-examine their organizational models, engaging the entire Army to ask, "Does our transformation still make sense? What in our organizational culture needs to be reinvented? Where are the soldiers who are getting out going and why?" The biggest problem with such an undertaking is that the most powerful individuals stand to lose the most in an organizational transformation. Often, they will resist the changes necessary to move ahead in order to avoid losing power.[3]

Finally, there is the issue of Army culture. In any change effort, culture plays a vital role, either as a facilitator or a barrier. Leaders must learn to harness the positive dimensions of a culture in the change efforts. Culture is much like the water in an aquarium. While it is largely invisible, its chemistry and life-supporting qualities profoundly affect its inhabitants. An Army requiring transformation is like an aquarium polluted by too many algae, and requires restoration of the right balance of elements in the environment. Leaders affecting an Army transformation must understand the vital role the corporate culture plays in any change efforts.[4]

Today's Army Culture

> "... processes and structures that lend required order and routine to our lives can also hinder innovation. Examples include human resource policies that manage people as inputs rather than outputs, labyrinthine organizational structures that frustrate interdisciplinary networking, and reporting procedures that focus more on things then on ideas."
>
> BG David A. Fastabend and Mr. Robert H. Simpson, "Adapt or Die"
> *The Imperative for a Culture of Innovation in the United States Army*[5]

The Army's (as well as military's) cultural barriers will be the most potent speed bumps to reform of the Army and this impacts recruiting and retention. General George Patton observed that armies "tend to consider the most recent past war as the last word, the sealed future pattern of all contests we realize, none better, that in the last war it was necessary to make many improvisations and to ply our trade with ill-assorted tools."[6] The Army is adapting in Iraq under selective command climates, but how about the culture? What happens when the very factors mentioned in the 2001 Army Training, Leadership, and Development Panel (ATLDP), or the "Blue-Ribbon Panel," such as "Micromanagement has become part of the Army culture," remain in place in the Army that is in the continental United States as veterans return? Will those who lead the Army, at all levels, promoted under the current culture, modify their behaviors and climates to those who have adapted?[7]

The ATLDP was sanctioned in 2000–2001 by Army Chief of Staff General Eric Shinseki in reaction to the results of negative officer and enlisted surveys as well as in light of the transformation of the Army and the emerging operating environment. As part of the transformation process, the panel was asked to identify the characteristics and skills required for leaders of the transforming force. General Shinseki also tasked the panel to examine the current systems for training and leader development, but most importantly, he asked the panel to examine the Army culture.

During the 1990s, the rate at which junior officers, primarily captains, left the Army nearly doubled, from 8 percent (1,497 of 16,933 captains) in 1996 to 14 percent (2,208 of 15,404 captains) by 2000.[8] In 2007–2008 there are also signs of an exodus of junior officers believing they are seen but not heard and this amounted to a shortfall of 1,474 captains and majors.[9] This has been highlighted by the first time ever offering by the Army of $20,000 to captains who will remain on active duty.[10] The primary cause behind both exoduses was not the economy as was forecasted, or continual back-to-back deployments, but a frustration with micromanagement and the Industrial Age training system that had evolved from the times of Frederick Taylor in the late 1800s and that left no room for innovation. Retired Army Colonel Don Snider, a professor at the U.S. Military Academy at West Point, NY, says, "They did not sign up to become bureaucrats, but you become one when you cannot exercise any discretionary authority."[11]

The review's Training and Leader Development Panel published findings in May 2001 based on interviews with 13,500 active-duty and reserve officers. Junior officers complained that they were micromanaged and discouraged from learning through their own mistakes and experiences. Perhaps most damning, many believed they had no future in the Army. And they had cause to be concerned, with many officers being asked to leave as a result of one less-than-glowing officer evaluation report. Many different complicated factors culminated into an Army culture of caution, causing promotion anxiety and careerism. The findings were discovered again when Brigadier General Michael S. Linnington was tasked by the Army Chief Staff General Casey to discover the causes of officer exodus from the Army in 2005–2007.[12]

The result of this culture was that leaders at all levels were afraid of failing. Again, the impact of the Army culture on its performance in Iraq was recently reflected in an article, "A Failure in Generalship: The Crisis in America's Senior Officer Corps," by Lieutenant Colonel Paul Yingling in May 2007.[13] Yingling's article became the subject of meetings and discussions throughout the Army. Its underlined thesis was that the Army's senior leadership has not evolved as war has evolved, and its personnel system is picking the wrong people to lead the Army in future conflicts. Greg Jaffe summarizes Yingling's point:

> In his controversial essay, Col. Yingling pinned much of the Army's failings in Iraq on generals who he says didn't prepare for guerrilla fights in the decade prior to the war, and then didn't adjust as quickly as front-line troops. Young officers had to adapt to survive, he wrote. The generals, products of a system that encouraged conformity and discouraged risk takers, were often a step behind the enemy, he said. "It is unreasonable to expect that an officer who spends 25 years conforming to institutional expectations will emerge as an innovator," he wrote. The solution, he said, is to change the way the Army selects and promotes generals, taking into account reviews by subordinates.[14]

The Army has begun to address these cultural issues, while not publicly announcing them. This has led to the evolution to today's Basic Officer Leader Course (BOLC). The report found that the curricula of most of the courses comprising the officer education system emphasized technical skills that taught officers "what to think" over "how to think." The panel also said that lieutenants needed platoon leadership skills, an understanding of how their platoons fit into the larger picture (impact), and more hands-on training.[15]

Previous studies support Yingling's recent assertion. For example, retired Army Lieutenant Colonel Leonard Wong published a report, "Stifling Innovation? Developing Tomorrow's Leaders Today," that confirmed the panel's findings. He interviewed company commanders. In one widely cited passage, a commander said, "They are giving me the egg and telling me how to suck on it." Shinseki backed the panel's recommendations. But as the Army began planning changes, Al Qaeda struck the World Trade Center's twin towers and the Pentagon in the United States. Overhauling the culture would wait as the Army went to war.[16] But did and will war dismantle the culture that the ATLDP and the recent 2008 study by BG Linnington find existed? Some insights say no, while conflicts with command climates that are encouraging adaptation and innovation will begin to occur as leaders and units rotate home from the war.

Sean Naylor's recent book *Not a Good Day to Die: The Untold Story of Operation Anaconda* illustrates what the soldiers went through operating under a culture that is still stuck in the Industrial Age. This culture includes an obsession with technology, which led senior commanders to believe that they could lead the campaign from thousands of miles away; operating with parallel chains of command that did not talk to each other. It is also a culture that believed that precision fires could compensate for a lack of "boots on the ground"—that the enemy would be arrayed in perfect templates for staff officers to plot and target them. Instead, the enemy fought and fought hard. In the end, the U.S. soldiers won Anaconda, but how much longer can we (the U.S. military) get away with this? Cultural evolution is generational.[17]

Don Snider argues that this is largely due to the Army's "control culture" as "its central elements derive from an attempt to deal with (and if possible overcome) the uncertainty of war, to impose some pattern on war, to control war's outcome, and to invest war with meaning and significance." This can be traced back to the evolution of modern Army to the cult of management science, from which the Army's personnel system has branched and developed as the foundation for today's culture. It must be this foundation that is broken up, removed, and rebuilt into a new foundation to support the new Army. But there are going to be many obstacles to this process.[18]

Compounding cultural obstacles include a combat-masculine-warrior "enculturation," exclusive policies, separatist attitudes, hostile interactions, and strong subcultures, such as the individual branches, which guard their "turf" stubbornly, many times undermining the goals of strategic leaders. The results are "bureaucratic forces" with authoritarian leaders with poor internal communications, who ignore possibilities and innovation.[19]

A fundamental lack of trust between superiors and subordinates necessitates increased supervision, setting the stage for micromanagement. Distrust of subordinates hinders delegation, thereby increasing the amount of information processed at the strategic level. This added burden slows down the strategic decision cycle. Meanwhile, the increased demand for decision making forces strategic commanders to legislate explicit directions (rules and regulations). As the volume of regulations heaped upon subordinates increases, the overall organization becomes a more rigid mechanical and bureaucratic whole.

"Method-driven orders" (evolving from increased regulation) increase the need for explicit communication between superiors and subordinates. A greater overlap of responsibility occurs, leading to confusion, friction, and competition for authority between superiors and subordinates. Mistrust ensures that tactics are mandated from one level down to the next.[20]

While not occurring with some leaders and units in Iraq and Afghanistan, tactical expertise can be underdeveloped. Subordinates are treated less like "thinking beings" and more like slavish robotic servants. The creative and intellectual ability of individual commanders is not used as much as it should be, generating dissatisfaction, cynicism, frustration, and low morale. Tactical commanders (given narrow authority) become hesitant to make decisions. The continuous need to obtain permission to execute plans stifles initiative. As they can no longer fluidly adapt to changing circumstances, they become trapped in rigidly defined roles, like clogs in a machine, and respond ineffectively to a rapidly changing enemy.[21]

As Don Snider rightly points out, the "control culture" runs deep. To a degree, the control-centric culture is necessary to prevent large standing forces from becoming "engines of despotism." The art of the process is to apply that control of the military writ large while maintaining the organizational flexibility to ensure it is an effective force. That's no easy trick.[22]

One of the promises of complexity science is the notion that small interventions can have massive effect when amplified over time. Therein lies hope for Army evolution. There is just no guarantee that the interventions made will have a desired effect. There are multiple levers of change and it is no surprise that I consider the educational system a very important one, especially as a culture-embedding mechanism. Even more important are the incentive structures (awards, rewards, punishments) that every member of the organization can see each day. If culture change can be accelerated, it will be through those highly visible mechanisms.[23]

There is good news, though, in that an example exists of how "why" and "what" are evolving parallel with "how-to" in Army basic training and advanced individual training across all military occupational specialties. The evolution that TRADOC is creating in how we train and prepare soldiers for 4GW and how they are doing it through the leadership of drill sergeants—actually leading new soldiers as a squad through complex small unit training such as convoy operations and reaction to individual explosive devices—are a small piece of cultural evolution through parallel evolution. This is the end result of the filtering down of the vision moving forward toward the goal.[24]

But this understanding, with a demonstration that the Army can put to execution a small "how-to" of parallel evolution through its preparation of new soldiers, must also be applied across the board. The evolution that began with enlisted basic and AIT now must evolve to include the way the Army prepares potential leaders to become officers in a revised accessions process. The new development, the BOLC, is a dramatic departure from how the Army has prepared in the past, but as BOLC begins to turn out the newest officers based on lessons learned during war, a culture must be in place to take care of and support these individuals.[25]

The Culture We Say We Want

The importance of changing the Army's culture is a significant—arguably, the most significant—component of the Army's ongoing transformation efforts. The relevance of the discussion and emphasis on culture are significant in the Army today for four primary reasons: the strategic environment, ongoing transformation, operational environment, and first time for the All Volunteer Force.[26]

Unlike many of the Army's previous transformation efforts, transformation of the Army is occurring in a compressed timeline, and while it is at war. That translates to an almost unprecedented strain on the force. Undergoing changes in culture is an inevitable by-product of being at war for a long duration and also of any large-scale transformation, which ultimately creates many intended as well as unintended consequences. Collectively, such changes represent another factor for soldiers of all ranks to ponder as they contemplate the duration of their commitment to serving in the Army.

Also, the Army's historic tendency to focus internally and to not openly tell the Army's story to a myriad of external audiences may be a fatal shortcoming in the Information Age, where information flows instantaneously, and the competition to attract quality people is so strong among colleges and universities, the other armed services, and private sector companies, especially as the global war on terror continues.

In order to field soldiers and units prepared to defeat today's and tomorrow's enemies, the Army is in the midst of its most significant transformation since the period after the Spanish-American War under Secretary of War Elihu Root (1899–1904). Army transformation is not only impacting its units, such as turning Cold War–era divisions into twenty-first century brigade combat teams (BCTs); it also involves an individual transformation of sorts by inculcating all soldiers, regardless of their rank and military occupational specialty (MOS), with the warrior ethos and a "soldier first—specialist second" mindset. These organizational and individual transformations represent a change in culture, if for no other reason then because they represent a significant departure from long-standing practices.

Establishing and practicing *trust* involves more than rhetoric. A strong common outlook between superiors and subordinates establishes mutual trust. This reduces the need for continuous supervision and thereby diminishes micromanagement. Strong trust encourages delegation and reduces the amount of information and tactical direction processed at the strategic level. With less information to process and a greater focus on strategic issues, the strategic decision cycle accelerates and the need for legislative direction (rules and regulations) diminishes, creating a more fluid, organic whole.[27]

Objective-driven orders (those that emphasize the end-state of operations over methods) reduce the need for explicit communication between superiors and subordinates. Consequently, little overlap of responsibility occurs and a good economy of effort emerges that expands the organization's overall ability to influence the environment.

Tactics are not mandated from above but are developed and implemented at each level by those closest to the problems at hand. Thereby, the creative ability and intellectual capacity of each individual commander is harnessed, resulting in greater morale and enthusiasm. All of these factors combine to increase the probability of finding effective solutions.

Tactical commanders (given broad authority) take the initiative and become semi-autonomous; they continuously interact with the environment without hesitation. This speeding up of decision cycles among tactical commanders provides quick adaptation to the changing environment. Collectively, tactical adaptation works up the chain of command to influence the evolution of the organization as a whole.

Because of the nature of the GWOT and the manner and locations in which we are now most actively waging the war—such as Iraq and Afghanistan—in order to be effective, the Army must have a better appreciation for the customs, courtesies, and cultures of those and other nations where the United States is executing its global commitments. This requires, or should require, the Army to re-think its own service culture in order to optimize its capabilities in the operational environment.

The ongoing U.S. military operations in Iraq and Afghanistan represent the first time the United States has been involved in protracted combat operations since the inception of the AVF in 1973. Conducting protracted combat operations in which soldiers are dying and being wounded on a daily basis not only affects retaining soldiers currently serving in the Army, but also recruiting American men and women into the Army.[28]

Changes in the strategic environment since September 11, 2001, necessitate changing Army retention and recruiting "culture" because the same advertising strategies, master messages, target-audience analyses, influencer outreach efforts, and so on that were used before 9/11 will no longer be successful without being updated to reflect the realities of today's strategic landscape. Without changing other factors, just adding more recruiters and more retention of noncommissioned officers (NCOs), and more money to the Army's recruiting budget, likely will not enable the Army to reach its recruiting and retention goals in the active or reserve component.[29]

Any study of U.S. military culture reveals an interesting paradox. Change is endemic to military culture; for example, social changes include racial integration in the 1940s and 1950s, changing to an AVF in 1973, expanded opportunities for women in the 1980s and continuing today, and evolving views on homosexual service members. Changes also occur in how the Army fights, particularly in force structures and doctrine such as Marshall's reforms of the Army 1939–1943 and the renaissance with Air-Land Battle in the late 1970s and through the 1980s.[30]

People in the military are also accustomed to missions changing on short notice, being reassigned frequently during their careers, and working with and for new people all the time. But while change is a way of life within the military, so is inertia; people at all levels resist change—in equipment; organizational design; uniform accessories; long-established and accepted policies; tactics, techniques, and procedures (TTP); and so on. This inertia is not unique to the military; human beings are fundamentally resistant to change. It is almost always easier to maintain the status quo.

Since the Army is a people-centric organization composed of volunteers, the Army is always in the retention and recruiting business. Because of that constant need to sustain the Army as a force with committed professionals, the Army can't simply be reactive and let cultural changes within the Army occur solely as a by-product of other changes. The Army must be proactive and lead the cultural aspects of the profession of arms as the Army does with everything else that matters to its soldiers. Cultural change in a profession where the Army's Constitutional charter is to win the nation's wars, and where soldiers must be prepared to make the ultimate sacrifice, is fundamentally a leadership challenge and leaders must lead the change.

The Army must change its culture to meet the realities of the twenty-first century and the post–September 11, 2001, world in which the people of the Army live and operate. Merely transforming the Army's organizational structure and design, and how the Army trains individually and collectively, will not, in themselves, produce the types of changes in Army culture that will serve the institution well over the long haul.

The strategic environment—both the environment in which Army units and soldiers are operating as well as the domestic environment from which we recruit—is significantly different today from what it was only a few years ago. Many things that worked previously will no longer work, at least not without modification, including the ways the Army sustains its AVF. Merely leaving the cultural aspects of our ongoing transformation to chance will not produce optimum results; it will be neither effective nor efficient. Through deliberate—not bureaucratic—thought, and by bringing the right people together, I believe it is possible for strategic leaders to lead the Army's cultural transformation the way they are leading our ongoing organizational and training transformation. Leadership is key, particularly leadership that can break from the past to move the Army into the future.

John P. Kotter's *Leading Change* tells us that quality leadership in transforming business (large) organizations is the key to successful change. Over the course of 25 years, Kotter studied hundreds of businesses and identified eight common errors that prevent organizations from accomplishing their change objectives. Kotter has an eight-step process for any organization to follow to avoid his common errors and to succeed.[31]

Kotter's model provides a good foundation to compare with the *parallel evolution* model. He discusses the significant changes that business organizations underwent to keep pace with their changing operating environment (competitors).

Kotter states that change is hard, and he believes that the challenges are avoidable. Kotter's eight-step process is

1. Establish a sense of urgency.
2. Create the guiding coalition.
3. Develop a vision and strategy.
4. Communicate the change vision.
5. Empower employees for broad-based action.
6. Generate short-term wins.
7. Consolidate gains and generate more change.
8. Anchor new approaches in the culture.[32]

Kotter argues throughout this eight-stage process that quality leadership is critical. Kotter states that only leaders can cut through layers of negative corporate resistance to motivate people to change their behavior. This makes a sharp distinction between managers and leaders. Only leaders can successfully "anchor" change in a business culture. Simply "managing change" is not enough. True transformation comes about, according to Kotter, only through "leading change."[33]

Kotter provides a model for organizational change within corporate businesses that may have value for Army leaders. How applicable are the principles of Kotter's *Leading Change* model for transforming modern businesses to today's transforming Army? Together, Kotter's theoretical model and historical examples of parallel evolution in this chapter provide today's Army leaders with "lessons learned" on how to go about evolving the culture—changing the Army. Kotter's eight-step model is the best lead-in for a comparison to the introduction of parallel evolution: the changing of a military organization to make it effective as war evolves. But as Kotter emphasizes, and both successful and unsuccessful examples allude, leadership is key.

Given that overseeing and managing change are important, the bureaucracy was established and maintained to keep and track the cost of such efforts, many times to the determent of such efforts as the organization focuses inwardly and process is sustained at the cost of the outward quest for an improved organization. But for the Army, the much bigger challenge is leading change. Only leadership can blast through the many sources of cultural inertia caused by the personnel system's measure of success within the Army. Only leadership can motivate the actions needed to alter behavior in any significant way. Only leadership can get change to stick, by anchoring it in the very culture of the organization. But this is not confined to only the strategic leaders.[34]

Leadership, Kotter stresses, cannot be confined to one larger-than-life individual who charms thousands into being obedient followers. The Army is far too complex to be transformed by a single giant. The leadership effort must have support from many people, at all levels, each of whom assists the leadership agenda within their sphere of activity.[35]

Looking at classic examples of transformation from the Army's own history within the context of Kotter's model also serves to highlight the degree to which these military case studies mirror the theory.[36] This chapter is followed by three chapters on how to shape the strategic environment to allow an implementation of reforms to create and nurture adaptive leaders.[37]

What Is Parallel Evolution?

> "Culture evolution occurs (changes) one retirement and firing at a time."
>
> Jack Welch[38]

The Army, particularly senior leaders, must take time to understand previous attempts at changing culture, within the Army and by others (both inside and outside the military). Which worked well, and why? Failures can point to subtle systemic barriers that need to be overcome or bypassed. An improved understanding of all the building blocks—strong leadership, understanding the organization, building the case for change—provides a sound foundation for the second stage, developing a strategic plan. Without a thorough understanding, the wrong strategy may be selected.[39]

After development of a vision defined by "what" and "why," then comes a strategic "how-to" plan to achieve *parallel evolution*. Implementation of this plan is the hardest part of evolving a culture. The first step is building a platform, and Army Chief of Staff General John Casey has to do a good job with making the case so that soldiers can see what is in it for them. Second is communicating the change, which has to be done in light of the ongoing war. The hard part is executing cultural evolution to nurture desired future traits, many that can stand in contradiction to measures of success in the current culture.

Strategic leaders must reinforce the evolution toward the end state—in the case of the Army, moving from a hierarchal, top-down, mobilization-based doctrine Army to an expeditionary Army that is flattened, seeks constant adaptability, and rewards productive innovation, as long as the Army is out-thinking its current enemies. They do this with structured incentives, focusing on the people aspects of change, providing opportunities to experience the change first-hand, and developing an infrastructure for consolidating the desired culture (e.g., a center of expertise without regard to its members being considered successful in light of the old culture). Success of any change initiative depends on implementation. This is the hard part. This is a systematic process of discussing how and what, questioning, following through, and ensuring accountability. Support from the top is crucial to facilitate all other changes. The examination of second- and third-order effects is a must, and was not done with the Cohesion Readiness and Training (COHORT) system in the 1980s.[40]

The Army must build its capacity for culture change. Capacity building involves providing incentives to recognize and reinforce new behaviors. In the process, it is important to focus on people and ensure that strategies are in place for dealing with the stress of change. One strategy is to provide learning experiences for soldiers and leaders, allowing them to try out the new behaviors during the transition period. Finally, deal with gaps in infrastructure by identifying and providing tools and training to get the job done.[41]

The U.S. Army has the opportunity to achieve parallel evolution. Many examples exist within the U.S. Army, but examples also exist in the armies of other nations, such as the Prussian Army and the Israeli Army. The best and closest the U.S. Army came to achieving *parallel evolution* were its attempts after Vietnam. While the Israeli Army provides several lessons, it essentially evolved from a very small force into a dominating Army, and never experienced change "mid-stream," so to speak, as did the Germans amid World War I and as the United States is trying now, as it fights.[42]

The Prussian Army and the U.S. Army are two historic case studies that illuminate the role of institutional intellectualism in producing parallel evolution. The first provides an example of a specially constituted team of intellectuals responsible for transforming an entire military organization in response to an adversary's military revolution: the Prussian reforms followed catastrophic defeat to Napoleon at Jena-Auerstädt in 1806. The second example demonstrates that reforms must be complete in order to move the organization toward the goal of operating and winning in a future operating environment.[43]

The U.S. Army's major reform was its reaction to the Vietnam War. This dramatic reform in the 1970s and 1980s affected almost every aspect of the Army—except the officer personnel system. While attempts were made at unit manning, the Army failed to address factors that impacted the development of an officer's character. The Army won the first Gulf War, and is adapting in Iraq; but the Army's strategic leaders are often unaware of the influences that have shaped their cultures. The Army must identify the state of readiness for change, including potential barriers, so that an agenda for change can be pieced together.[44]

Parallel Evolution, Historical Examples

The German (Prussian) Army from 1809 to 1942 provides historical insights for "how-to" evolve a culture using *parallel evolution*.[45]

Between 1919 and 1933, Chief of the German General Staff General von Seeckt oversaw a strategic "how-to" based on a critical analysis of German Army performance in World War I. The question was, how did the Germans lose despite the success in both the tactical defense (changes made in 1916), and the tactical offense (changes made between 1916 and 1918)? More so, how could the Germans capitalize on the success of the tactical offense and move beyond the battlefield to the operational depths to attack an enemy's vulnerabilities before the enemy could relocate reinforcements?[46]

The Germans brought together officers of all ranks, but mainly captains and majors, experts in an array of fields, and had them write a report on their area of expertise as it pertained to German performance in World War I. After von Seeckt left, German military leaders continued to nurture the Reichswehr, a new military organization built on the wreckage of the old Imperial Army.[47]

Yet, at the heart of the parallel evolution of the old Imperial Army was a culture that encouraged innovation and strength of character. The evolution of the culture began in 1801 under the leadership of Gerhard Scharnhorst, with key members in place to pick up the pieces after the disaster of Jena-Auerstädt in 1806. The members of Scharnhorst's military society had written papers and presented them to other members of the group dealing with aspects of the old Army under Frederick the Great that had to be reformed in light of the French Revolution.

Scharnhorst recognized the scope and implications of the new form of warfare arising from the French Revolution. He responded to this event by creating a military society for investigating and discussing military history, strategy, tactics, and current developments. That is, he perceived a need to plan for the future and provide a base for looking ahead. He realized that soldiers must study war in the context of the social, political, economic, technological, and moral factors that influence military institutions and operations.[48]

He also advocated creation of an institutionalized genius within the army by establishing a general staff system in which staff officers worked with the commander and provided inputs. Scharnhorst recognized that an army would not always have a genius like Frederick the Great or Napoleon but that educated and capable officers serving in staff positions and advising a field commander could assure such genius on a continuing basis. He envisioned staff officers rotating between staff and field positions, providing a free and steady flow of ideas throughout the general staff and the army itself. After its defeat by Napoleon in 1806, Prussia began Scharnhorst's reforms in earnest.[49]

In his discussion of Scharnhorst, Carl von Clausewitz noted that, in changing times, people are comfortable building systems or boxes into which they fit the world: "It was to Scharnhorst's great credit that he was not in the least influenced by these pretentious theories [systems], which at the time were over whelming everyone who did not unthinkingly cling to the past. He recognized both the unchanging elements in the present age and the inadequacy of old methods, but he wanted the new to emerge from the old, and he wanted to arrive quickly with as little fanfare as possible at a new, appropriate method." Basically, Scharnhorst avoided rushing to determine the boundaries of the times, choosing to study events and allow the change to define itself.[50]

Thus, Scharnhorst laid the foundation of the general staff and provided the vision for its future growth. The key ingredient in his approach was belief in the potential of people with talent and ability. To him, the most crucial element in analyzing and dealing with change was people and the way they respond. The significance of these reforms easily could go unnoticed.

The proof of their efficacy lies in the results of the Battle of Waterloo; in his scholarly work, *The Enlightened Soldier,* Dr. Charles White credits Scharnhorst's reforms for Napoleon's defeat in that battle.[51] Specifically, August Wilhelm

Gneisenau, Field Marshal Gebhard von Blücher's chief of staff, recommended that von Blücher return to the battlefield—an action that caught Napoleon off guard before he could defeat Wellington. These reforms marked the beginning of the Prussian/German general staff system, which, according to Trevor Dupuy, institutionalized excellence within the military.[52]

Scharnhorst and his students saw that human affairs are dynamic and that technology changes. Thus, weapons, equipment, and armies must change with the times. This called for an environment to foster continuous improvement. The Army found the key to German military effectiveness and how they evolved their culture in events of the early nineteenth century.[53]

In "Revolutions in Military Affairs," Commanders James R. Fitzsimonds and Jan M. Van Tol recognize the German Army for its development of moving tactical Sturm tactics (later termed *blitzkrieg* by the English) to the operational level between the world wars, noting how the Army was open to new technologies in warfare. However, the authors missed the reason for the Germans' ability to recognize the factors at work at that time and to use those factors to their advantage. It was Scharnhorst who created an environment in which soldiers were encouraged to be innovative and "think outside the lines." This laid the foundation for the abilities of von Seeckt.[54]

Moreover, Scharnhorst's reforms were especially significant because they were not a formalized doctrine but constantly evolved, based on changes in technology, doctrine, geopolitics, and anything else that might affect the employment of military power. It was this intellectual ferment that laid the base for the transformation of the German Army from tactical maneuver warfare to operational maneuver warfare between the world wars.[55]

The crucial transformations in German military tactical doctrine, organization, and training laid the foundations for fighting Germany's future wars. Von Seeckt's influence demonstrates how the general, along with a few other "visionary" officers—including armor tactician Ernst Volckheim and air tactician Helmut Wilberg—collaborated to develop the core doctrine for what became operational-level Sturm tactics or blitzkreig.[56]

The Germans wrote an After Action Review in 1919 that provided a blueprint for change, with the goal of the Germany Army operating in smaller but highly mobile, high-quality forces that would be made even more efficient by support from aircraft. The vision of von Seeckt's plan was carried over even after he left the position in 1927. In 1929, a German signal officer and general staff officer, Heinz Guderian, was assigned to the common German-Russian tank school in Kazan, which was established to allow the Germans to bypass the post-war limitations by training in operating tanks in Russia.

Guderian read the British books about the new principles of tank warfare, and although Lidell-Hart's ideas were given little military attention in Britain, Guderian practically developed them over several years and demonstrated them to his superiors in military exercises and war games. (Britain would provide a good example of not how to do *parallel evolution* because it did not evolve its culture.)

Two factors helped Guderian's tactical ideas become dominant in the German Army more than in any other Army. The first factor, as mentioned in the Introduction to this book regarding the culture of a Third Generation Warfare Army and early in this section, was an Army culture that expected German officers to question old ways and to examine new ideas. The second factor was that after World War I, the German Army was only allowed to have 100,000 personnel, including fewer than 5 percent officers. So, a strenuous process was used to pick the best and the brightest, while NCOs were selected based on their ability to become officers in time of mobilization.[57]

So von Seeckt began a transformation with a culture that already expected ferment and people who could make it happen. Most importantly, it allowed von Seeckt, with the assistance of many like-minded subordinates, to lay out a "how-to" plan to implement and drive the desired changes the German Army sought:

> The lesson that von Seeckt holds for us in the 1990s is not that of the developer of blitzkrieg (Corum accurately points out that this type of war awaited further development but that the combined arms concept that was at the root of blitzkrieg was a key goal and achievement of von Seeckt) but rather as one who saw the comprehensive nature of that which we call military art and science.
>
> Using this language of "comprehensiveness" today might well summon up images of the great "jointness" debates in the US, but von Seeckt went well beyond that type of analysis. Rather, his contribution was to offer an example of military reform from alpha to omega, the whole series of connected actions that constitute a meaningful reforming of a military. Thus, it is important to have a good doctrine, but just as important is the rather mundane question of how one actually implements the new thinking. As Corum points out, "Sound tactical theory aside, it was in training that the Reichswehr surpassed all its contemporary rivals, ensuring the battlefield efficiency and tactical success of the German Army in 1939/40." (p. xvi) He shows in the text, chapter by chapter, how von Seeckt left his Berlin headquarters that had so well developed the new operational doctrines, went to the field and demanded again and again that the troops implement the new concepts of mobile warfare. We might add that among these new concepts was "air-land battle," the lessons about which the air sections of the Truppenamt learned from their World War I experience and then codified under von Seeckt's direction. Additionally, Corum emphasizes the fact that the best field training and the best doctrinal development is useless unless the link between the two, the middle-level officers, is of such competence that the doctrine can be executed. This competence did not come about due to traditional German efficiency (although it may have been aided by it). It was built brick by brick by von Seeckt, year by year, using many avenues of officer education and training, many of which he personally oversaw.[58]

Today the Army must ask itself if it has such an officer corps. The Army has gone through the throes of addressing the questions of balance between "generalists versus specialists" or of teaching "operations" versus "strategy." But does the

Army encourage mid-level officers such as Volckheim and Wilberg, who did not produce large-scale theory but did produce the link between theory and practice that von Seeckt's reformation demanded?

Further, for every such thinker, there must be dozens of thinking Army officers off whom the thinker can bounce ideas to see, day by day, how the new concepts must be put into practice. One would have to look in vain for such officers as they return from Iraq, or the ones who led the pleas for change prior to the war in Iraq, but who were retired. The Army has them; they now have to have the moral courage to elevate them to help bridge the cries for change and the goal the Army says it wants to reach.

The U.S. Army's Renaissance of the 1980s—Trying to Achieve Parallel Evolution

Historians usually discuss military history in terms of battlefield success. This is called "results driven analysis," which many times leads to false examinations of history. A more difficult but more thorough and exact way to use history is to examine it from the perspective of the influences that impacted the decision. Very few focus on the day-to-day decisions and peacetime victories that shape the military organizations that fight those battles.[59]

The Army change after Vietnam in the late 1970s and 1980s is another example of attempts at *parallel evolution*. An incredible metamorphosis of the American Army took place between defeated pariahs in 1972 to victory in 1991 in the Gulf War. One must start with Vietnam-era catastrophes, move on to the "dark ages" of the 1970s, which followed that era, and analyze the policy and doctrinal changes that carried the Army to the end of the 1980s, in light of the tactical success of Desert Storm. The strategic leaders of today must learn from the experiences of the past generation with the Army's best attempt at *parallel evolution*.[60]

It is the attempts at evolving the culture from Vietnam to the Gulf War that need to be compared and contrasted today. In Vietnam, the Army had a culture in which command and control failures flowed directly from the dissonance between doctrine and policy. The Pentagon kept a tight rein on the scope of the conflict, and General Westmoreland made company-level command decisions from Saigon and would not allow local command discretion.

Epidemic command and control failures effectively destroyed morale and integrity. The first casualty of a war that made no sense was integrity. Junior leaders struggled through a quagmire of moral ambiguity as they tried to comply with illogical orders from above without needlessly killing the people in their command. Staff officers developed the practice of reporting statistics "construed to show marked progress in a war where none really existed" in response to the distorted command emphasis on body counts.

Officers in the field learned to develop their own tactics to survive. Sometimes this meant directly disobeying orders. While massive amounts of resources allowed for and facilitated several evolutionary reforms, were the factors that undermined the Army in Vietnam changed? A new period must occur starting today, in the twenty-first century, that combines the best lessons on how to prepare future leaders to move the Army from its Industrial Age force (Second Generation) to a maneuver-based force (Third Generation) that can cope with what is called Fourth Generation Warfare (non-state war).

Several officers who had served as junior and middle grade officers in Vietnam were determined to reform the Army's culture of management science to one that focused on war fighting. Four significant events occurred during the 1970s and 1980s, together with the influx of material that enabled the Army to perform effectively in Panama and the Gulf War. Several of these efforts evolved together.[61]

Three of the events succeeded in preparing the Army to conduct warfare with a Soviet-style heavy force. These were (1) the post-1973 "doctrinal renaissance" begun by General William DePuy's advocacy of the doctrine of active defense and carried on by General Donn Starry with Army doctrine evolving into the Air-Land Battle; (2) the Army training revolution, which produced realistic force-on-force training for Army leaders, soldiers, and their units, with lessons from this training shaped by "rank-blind" and candid after action reviews; and (3) a force design evolving in parallel with technologies and doctrine that included a continual development of air-assault doctrine that began in 1963, and included the 1st Cavalry Division experiments in 1971–1974, the 9th Infantry Division Motorized, High Technology Test Bed, and the 7th Infantry Division (Light), Army of Excellence force structure.[62]

The final significant event was attempts at changing the personnel system in light of the AVF, and the attempt to switch from an individual to a unit replacement system. These endeavors met with mixed results. Limitations were placed upon the Army by the Defense Officer Personnel Management Act (DOPMA) of 1980 and its rigid management policies.

DOPMA standardized OPA 47 across the services, and continued to hold the Army to a structure that would allow it to mobilize in case of a large land war against the Soviet Union—meaning a larger-than-necessary officer corps bound by the "up or out" promotion system. Despite these restrictions, which were seen as the only way of doing business by much of the officer corps, policies were implemented at lower levels to circumvent the restrictive boundaries set by DOPMA, son of OPA. Despite these restrictions from "above," there are examples indicating a culture built upon trust and professionalism did exist.

General Edward "Shy" Meyer, Army chief of staff (1979–1983) and Lieutenant General Walter Ulmer, commander of III Corps, Fort Hood, moved to give officers more experience, while creating positive leadership environments. Command tours for commanders of battalions and brigades were extended up to three years. Furthermore, General Ulmer moved to create a positive command climate in his corps that broke the tradition of top-down leadership doctrine.[63]

A top-down command style had evolved since World War I, and had become the norm of the management science culture because of the Army's perceived success in World War II. Top-down or authoritarian leadership requires blind, instant obedience from subordinates who were to be ruled by fear, creates

"Darwinian" competition among subordinate commanders, doles out information to subordinates, lies freely to them, and makes them lie to the senior when the truth would make the senior uncomfortable or make him look bad to his boss. This style of leadership also never takes responsibility for setting priorities, so that when anything is left undone, the senior can always find a subordinate to blame ("I told you that was top priority"—just as is everything else).[64]

Some of these moves had lasting impacts, as the division commanders who went into Desert Shield/Storm had a wealth of command experience at the battalion and brigade levels. Others did not last, as General Meyer and his subordinates moved on to retirement. It seemed the culture waited them out, because soon after General Meyer left the office of chief of staff, the new chief of staff of the Army, General John Wickham, Jr., reversed policies that created cohesion and experience because of demands to maintain the tradition of equality that existed in personnel management. Soon the glow of victory and moral purity of driving the evil Saddam Hussien out of Kuwait would ironically wash away any reforms in the personnel arena achieved in the 1980s because the foundation they rested upon was not fixed, just covered up.

What is the U.S. Army doing today, amid two campaigns in Iraq and Afghanistan, to prepare for Fourth Generation Warfare?

"Campaign Quality Army with Joint and Expeditionary Capabilities"—Parallel Evolution?

Despite the pressures of current operations, in the past two years, the U.S. Army has begun a root-and-branch restructuring. It also introduced a new doctrine manual for fighting irregular warfare, "Field Manual (FM) 3-24 Counterinsurgency" (also printed as "Marine Corps Warfighting Publication" [MCWP] 3-33.5) in 2006 (the manual is unlike any attempt before it, written by some of the most brilliant minds on the subject of insurgency warfare in both the U.S. Army and Marine Corps). There are also actions taking place in personnel policies, training, and professional education, that, when all taken together, resemble *parallel evolution*. The goal of the U.S. Army is to move from an Army that practices centralized and top-down controlled attrition and fights linear war to an Army that is a "campaign quality Army with joint and expeditionary capabilities."[65]

The U.S. Army calls this transformation, and it has been doing it while conducting large-scale operations in Iraq and Afghanistan. It is more than reorganization. It involves the adaptation of a modular force structure, a fundamental culture change, significant relocation and rebalancing of its global forces, radical modernization of equipment, and a reassessment of its personnel system. The majority of these initiatives are being conducted now, as this book is being written, continuing until 2013.[66]

Transformation is often confused with modularization or modernization. The conversion to a modular force is certainly part of the transformation, as is the introduction of new capabilities and equipment, such as the Future Combat System (FCS). However, transformation extends far wider and deeper. It consists of six primary activities. These include Redesign the Organization, Redefine the Culture, Relieve the Stress, Realign the Force, Invest in the Future, and Optimize the Capabilities.

As mentioned previously, the old U.S. Army structure comprised a loose affiliation of 10 highly independent active divisions, each with its own bespoke structure, ethos, role, and (in many cases) equipment. The emerging organization, now seeing action in combat, is strictly modular, built around three basic types of combined arms BCTs—Heavy, Stryker, and Infantry. These organizations no longer are hard-wired into their parent divisions, but can be affiliated, in a plug-and-play sense, into any formation. They are supported by functional Combat Service and Combat Service Support brigades, again built on common modular structures. The National Guard and Reserves have also been reorganized to conform to the same model.

This amounts to the biggest change to the U.S. Army structure since the end of the draft. The transformation program will run until around 2011, but the bulk of the changes, and in particular the reorganization of the BCTs, will be largely complete prior to this book being published in 2008.

This change has three immediate effects. First, it has driven large-scale reorganizing and reductions in overhead, which has allowed the U.S. Army to generate more combat power from within existing resources. Second, the Army is better configured to sustain the sorts of extended routine commitments in which it is now involved. Third, and last, the Army is now in a position to absorb the introduction of the FCS, and the doctrinal and tactical changes that will flow from that, with a minimum of additional organizational change.

Former Army Chief of Staff General Peter Schoomaker insisted that the culture within the U.S. Army must change. It appears that the new Army chief of staff, General George Casey, will continue to push for a cultural evolution.[67] This includes embracing a more adaptive style of leadership, the selection of leaders, and the personnel system—the bedrock of today's culture and one that remains stuck in the Industrial Age.

A cultural evolution is underpinned by the warrior ethos and the soldier's creed. The aim is to ensure that leaders in the U.S. Army are capable of facing and tackling uncertainty on today's and future battlefields. Cultural awareness education—both generic and specific—has been introduced into training at every level.

The personnel management system is also being evolved. The aim of the Army is to stabilize soldiers' lives and careers, while at the same time building solid units that become more adaptive as they grow cohesively and master the basics of that unit. Soldiers are staying longer within units to increase combat readiness and cohesion, reduce turnover, and eliminate many repetitive training requirements. A commander, his battalion commanders, sub-unit commanders, and soldiers will all form one BCT at the beginning of a three-year Army Force Generation (ARFORGEN) cycle (discussed subsequently). The entire team will remain together until the end of the three-year period, at which time the whole team will change.[68]

Despite many great efforts, the U.S. Army has not yet managed to introduce this system, principally because of the operational tempo of current operations and a tour interval of only 12 months, and additionally the inertia of a personnel system that has had at its centerpiece for more than a century, the individual replacement system (IRS). Since 1899, the U.S. Army has tried unit manning 12 separate times, and failed. Some of the failures can be attributed to poor analysis and planning, but successful ones, such as the COHORT system introduced in the 1980s, failed because they were a smaller part of a larger personnel system that retained the IRS.[69] With IRS having become accepted by the Army culture as the only way of doing business for decades, moving to a more effective system, as the Army is trying to do, will remain slow.

As part of the rationalization of the military estate and the change in the operational environment, the U.S. Army is undertaking a huge realignment of its forces referred to as Infrastructure and Global Basing. Two programs are currently in progress: the Base Realignment and Closure and the Integrated Global Positioning and Basing Strategy. Tens of thousands of troops are moving as a result of these programs, many from forward bases in Europe and Korea back to the United States.

While the U.S. Army converts to the modular force mentioned earlier, the Army says the introduction of the FCS technologies will also modernize it. The first of 15 FCS-equipped BCTs will be formed in 2014. The remainder of the BCTs—equipped with M1 Abram tanks and Bradley armored fighting vehicles—will be modernized through the use of FCS technologies. These 16 FCS technologies and equipment that constitute the FCS will be introduced in the field once they are mature and ready to deploy. The first of these technological "spin outs" is already being deployed to Iraq.[70]

The Regular Army is known as the Active Component (AC), and the Army National Guard (ARNG) and Army Reserve are known as the Reserve Component (RC). The Army is rebalancing the AC and RC. The rebalancing is being conducted between and within the components. There will be 48 AC BCTs and 28 ARNG BCTs. In addition, there will be a further 21 AC and RC support brigades.

The aforementioned balancing of the AC and RC, as well as the move to modularity, stabilization, and future force modernization, are linked by the adoption of ARFORGEN, a force readiness cycle. This will move the Army away from the tiered readiness system developed after World War II and supportive of the IRS. It will move the U.S. Army to a cyclical readiness system, very similar to that of other armies, such as the British Army's Force Readiness Mechanism, which promotes unit cohesion and readiness. With these parallel moves, the U.S. Army will have 20 fully equipped and trained BCTs available for immediate deployment, with a further 22 BCTs available on surge.[71]

As will be mentioned in more detail in the next chapter, in January 2007, it was announced that the U.S. Army was ordered to grow by a further 35,000 personnel, at a rate of 7,000 a year, a serious impact on recruiting and training of this force. Details of this enhancement are still vague as the Department of Defense has not yet completed its assessment of the exact force structure required. However, it will certainly involve an additional six AC BCTs and enhanced Combat Support and Combat Service Support brigades. This will take the U.S. Army to a total of 76 AC and ARNG combat brigades. It is estimated that this will cost an additional $70 billion, with $18 billion to be spent on equipment.[72]

The U.S. Army is evolving to the current and projected operational environment at a high tempo and while at war, which, while difficult historically, is the best time to interject changes. While the U.S. Army is creating modular brigades and command control headquarters to better meet the combatant commanders' operational requirements, it remains to be seen whether the Army is willing or even able to do away with its hierarchical Industrial Age force structure and personnel system that will only continue to diminish the full effects of its other efforts as mentioned earlier. Modularity is one of the building blocks for a foundation for a "campaign quality Army with joint and expeditionary capabilities," but it alone cannot build the Army the nation needs in the future.

Conclusion

The Army must take time to understand previous attempts at reform with a critical eye, including that of other nations, in periods of time, but also in the context of the cultures of those nations. What worked or did not work in one nation may not apply, but selected lessons can be used. Which worked well, and why? Failures can point to subtle systemic barriers that need to be removed.

An improved understanding of all the building blocks—strong leadership, understanding the organization, building the case for change—provides a sound foundation for the second stage, developing a strategic plan for the "how-to" mentioned later in this chapter. Without a thorough understanding, a changes in the culture will not take place, and soldiers and leaders will become frustrated and cynical of the word from the top; trust will begin to erode. Trust is the most important fabric that holds an effective Army together. But evolving the culture while other institutions of the Army are also changing is difficult.

Transformation must begin with the mind of the soldier. The late USAF Colonel John Boyd referred to effective change as consisting of "people-ideas-hardware," in that order. Boyd would insist that people use their minds when fighting wars. The U.S. Army is trying to do this, but is fighting against a Department of Defense and society that have flipped Boyd's approach on its head. Instead, U.S. Department of Defense transformation has focused on hardware-ideas-people (in that order).

Notes

I would like to thank Bruce I. Gudmundsson, Dr. Chuck White, and William S. Lind for their influence on this chapter.

1. John Boyd, "A Discourse on Winning and Losing," unpublished briefing (Washington, DC: August 1987), pp. 5–7, available at http://www.d-n-i.net. I am also indebted to Colonel T. X. Hammes, William S. Lind, Franklin Spinney, Colonel G. I. Wilson, and Greg Wilcox for their insights into Fourth Generation Warfare.

2. David Jablonsky, *Why Is Strategy Difficult?* (Carlisle, PA: Strategic Studies Institute, U.S. Army War College, September 30, 1992).

3. Colonel Richard H. Mackey, Sr., *Translating Vision into Reality: The Role of the Strategic Leader*, Student Study Project (Carlisle, PA: U.S. Army War College, April 15, 1992).

4. Edgar H. Schein, *Organizational Culture and Leadership* (San Francisco, CA: Jossey-Bass, 1992).

5. Brigadier General David Fastabend and Mr. Robert Simpson, "Adapt or Die: The Imperative for a Culture of Innovation in the United States Army," *Army* (Arlington, VA: Association of the United States Army, February 2004), p. 2; known herein as "Adapt or Die."

6. Colonel Joseph C. Carter and Major Michael S. Finer, "Stonewall Jackson and George S. Patton: A Survey in Leadership," *Infantry* (Fort Benning, GA: U.S. Army Infantry Association, January–February 2004), p. 4.

7. Chief of Staff of the Army, "The Army Training, Leadership and Development Panel Officer Study Report to the Army" (Washington, DC: Department of the Army, 2001), p. OS-9.

8. Discussions with Mark Lewis, August 2002. Also see Mark Lewis, *Time to Regenerate: A XGen's Response to Dr. Wong's Monograph,* retrieved from http://www.d-n-i.net/fcs/lewis_gen-x.htm.

9. Brigadier General Michael Linnington, "An Assessment of Stress on the Army, 1 Dec 07–28 Feb 08" (Washington, DC, March 2008), slide 16.

10. Colonel J. B. Burton, "Command's Feedback on Officer Critical Skills Retention Bonus," downloaded from e-mails sent to author (Iraq, 1st Cavalry Division, May 24, 2007); this memorandum was forwarded to me by several individuals.

11. Discussions with Dr. Don Snider, via e-mail, October 2003.

12. BG Michael S. Linnington, "Initial feedback from Tiger Team's visit to 15-mo BOG returnees," unpublished paper (Washington, DC: U.S. Army, January 2008), pp. 1–2.

13. Paul Yingling, "A Failure in Generalship: The Crisis in America's Senior Officer Corps," *Armed Forces Journal* (Washington, DC: DefenseNews Media Group, May 2007), p. 17.

14. Greg Jaffe, "Critiques of Iraq Reveal Rifts among Army Officers," *Wall Street Journal* (Washington, DC: Dow Jones and Company), p. 1.

15. "The Army Training, Leadership and Development Panel Officer Study Report to the Army," p. OS-2 to 9.

16. Dr. Leonard Wong, "Stifling Innovation? Developing Tomorrow's Leaders Today," Strategic Studies Institute (Carlisle, PA: U.S. Army War College, April 2002), p. 23.

17. Sean Naylor, *Not a Good Day to Die: The Untold Story of Operation Anaconda* (New York: Berkley Hardcover, March 2005).

18. Dr. Don Snider, Major John Nagl, and Major Tony Pfaff, *Army Professionalism, the Military Ethic, and Officership in the 21st Century* (West Point, NY: The United States Military Academy, 1999), pp. 8–12.

19. Discussions with Colonel George Reed, April 14, 2005.

20. ENS Steven Mason (USN), "Positive and Negative Leadership Models," Innovative application of Boyd's organizational climate to leadership at all levels. Author discussions

with Dr. Steven Stewart, Mr. John Tillson, and cadets of the Georgetown ROTC Program, particularly those with prior service in the Army.

21. LTC Gregory Wilcox (USA, ret.), "4GW, OODA Loops, and Implications of the Iraqi Insurgency," as delivered at the 16th Strategy Conference, Army War College, April 12–14, 2005.

22. Don M. Snider and Gayle L. Watkins, eds., *The Future of the Army Profession* (Boston, MA: McGraw-Hill, 2002), pp. 5–6.

23. Discussions with Colonel George Reed, April 14, 2005.

24. General Kevin Byrnes, talk to Association of the United States Army breakfast, March 3, 2005.

25. Noel M. Tichy, *Managing Strategic Change: Technical, Political and Cultural Dynamics* (Dearborn, MI: Wiley and Sons, 1983), p. 220.

26. United States Army, "Transformation Road Map" (Washington, DC: Department of the Army, November 2003), pp. 1–4.

27. John Kotter, *Leading Change* (New Haven, CT: Harvard Business School Press, 1999), pp. 101–117.

28. Interview with Colonel Mike Gallocious, March 21, 2005.

29. Jay Bookman, "Ominously, Army Recruiting Tumbles," *Atlanta Journal-Constitution* (Atlanta, GA, May 9, 2005), p. 11. Trent Geyax, "Military Blues, but not Green," *Newsweek* (New York, May 16, 2005), p. 16.

30. James Kittfield, *Prodigal Soldiers* (New York: Simon & Schuster, 1999), pp. 230–240.

31. Kotter, *Leading Change*, p. 71.

32. John P. Kotter, "Winning at Change," *Leader to Leader*, vol. 10 (Fall 1998), pp. 27–33.

33. John P. Kotter, "The Power of Feelings," *Leader to Leader*, vol. 27 (Winter 2003), pp. 25–31.

34. Kurt Lang, "Military Organizations," in *Handbook of Organizations*, ed. James G. March, (Chicago: Rand McNally & Co., 1965), p. 843.

35. Kurt Lang, "Military Organizations," p. 857.

36. Combat Studies Institute, "History of Transformation," *Military Review*, vol. 80, no. 3, (Fort Leavenworth, KS: U.S. Army Command & General Staff College, May–June, 2005), pp. 17–29.

37. The author is indebted to Dr. Steven Stewart for enlightening and teaching me about the missing aspect of my study: A strategic framework to facilitate reform of how the Army creates adaptive leaders.

38. Treasury Board of Canada, "Changing Management Culture: Models and Strategies to Make it Happen," retrieved from http://www.tbs-sct.gc.ca/cmo_mfc/Toolkit2/GCC/cmc-PR_e.asp?printable=True.

39. Daryl Conner, *Managing at the Speed of Change* (New York: Random House, 1992).

40. U.S. Army, "Our Army at War—Relevant and Ready—Today and Tomorrow, a Game Plan for Advancing Army Objectives in FY 05 and Beyond: Thinking Strategically," (Washington, DC: U.S. Army, November 1, 2004), pp. 14–16; "Change Army Culture to Reflect New Realities." Discussions with Dr. Steven Stewart January through March 2005.

41. U.S. Army, "Our Army at War—Relevant and Ready—Today and Tomorrow, a Game Plan for Advancing Army Objectives in FY 05 and Beyond: Thinking Strategically," (Washington, DC: U.S. Army, November 1, 2004), p. 14–16; "Change Army Culture to Reflect New Realities."

42. This is actually unfair, because the very changes sought by the Israelis prevented them from fighting long wars, which they correctly identified as being their greatest strategic weakness. So each of their wars lasted only a few days to a couple of weeks, with a few years between each one to evolve further from the lessons learned in the last war.

43. Another historical example that could have been presented in the book is how the British evolved to fight the insurgency in Malaysia. LTC John Nagl wrote an excellent book to examine the evolution, called *Counterinsurgency Lessons from Malaysia and Vietnam: Eating Soup with a Knife* (New Haven, CT: Praeger, 1999). Nagl's book compared the experiences of the British in Malaya and the United States in Vietnam, and argued that "a key issue for a counterinsurgent army is to calibrate correctly the amount of lethal force necessary to do the job with the minimum amount of nasty, counterproductive side effects. Even if using force with restraint meant the mission would take more time or reduce the level of force protection, it was still an indispensable step: a successful counterinsurgency took care and patience." But more than that, Nagl points to having a culture that will allow an Army to adapt in order to defeat its enemies. The British did this, and the United States did not. To summarize Nagl's assessment of this evolution in this book would not do Nagl's work justice. Peter Maass, "The Historian Who Now Commands a Tank Unit in Iraq" (New York: *The New York Times*, January 1, 2004): "In Malaya in the 1950's, the British successfully suppressed a Communist revolt (comprised mostly of ethnic Chinese) by generally steering clear of excessive force and instituting a "hearts and minds" campaign to strip the insurgents of public sympathy. In Vietnam in the 1960's and 1970's, the United States military took a different approach and failed. The Americans resorted to indiscriminate firepower and showed little concern for its effect on the civilian population.

44. Stephen Peter Rosen, "New Ways of War: Understanding Military Innovation," *International Security*, vol. 13, no. 1 (Summer 1988), p. 167; Michael Meese, "The Process of Organizational Change," in *Maneuver Warfare: An Anthology*, Richard Hooker, ed. (Novato, CA: Presidio Press, 1993), p. 205.

45. I chose 1942 because, by this time, Adolph Hitler had effectively wrested control of most of the decisions of the Army at the strategic level, and was beginning to influence operational and even tactical decisions.

46. Notes taken at lecture given by Dr. Bruce I Gudmundsson to Duke University cadets, October 13, 1998. The question was not whether the Germans could break through the trenches, it was whether they could beat Allied reinforcements, who, with the use of railroads, would be in position and ready to take on tired and bled down German formations.

47. James S. Corum, *The Roots of Blitzkrieg: Hans von Seeckt and German Military Reform* (Lawrence, KS: University Press of Kansas, 1992), pp. 21–45.

48. Charles E. White, *The Enlightened Soldier: Scharnhorst and the* Militarische Gesellschaft *in Berlin, 1801–1805* (New York: Praeger Publishers, 1989), p. 57. Alfred G. Gardiner referred to Scharnhorst's work as having "struck the keynote of scientific warfare," by which he meant that Scharnhorst saw the need to study war as a science and in connection with other changes taking place. "German Generalship," *Atlantic Monthly*, May 1916, p. 677.

49. T. N. Dupuy, *A Genius for War: The German Army and General Staff, 1807–1945* (Fairfax, VA: Hero Books, 1984), p. 18; and Omer Bartov, "The Nation in Arms: Germany and France, 1789–1939," *History Today*, September 1994, p. 29.

50. Carl von Clausewitz, *Historical and Political Writings*, Peter Paret and Daniel Moran, eds. and trans. (Princeton, NJ: Princeton University Press, 1992), p. 103.

51. White, *Enlightened Soldier*, p. 168.
52. Dupuy, *Genius for War*, pp. 300–307.
53. Walter Gorlitz, *History of the German General Staff, 1657–1945*, Brian Battershaw, trans. (Westport, CT: Greenwood Press, 1975), p. 8.
54. Commander James R. Fitzsimonds and Commander Jan M. Van Tol, "Revolutions in Military Affairs," *Joint Force Quarterly* (Washington, DC: National Defense University, Spring 1994), pp. 24–25.
55. Discussions with Dr. Charles White, professor, American Military University, October–November 1996.
56. Robert Corum, *Roots of Blitzkrieg* (Lawrence, KS: University of Kansas Press, 1992), p. xvi.
57. Robert Corum, *Roots of Blitzkrieg*, pp. 46–60.
58. LTC Douglas Erwin (USAF), "Review of the Roots of Blitzkrieg," *Airchronicles* (Maxwell, AL: U.S. Air Force Command and General Staff College, 1994), p. 32.
59. Lewis Sorley, *Thunderbolt: General Creighton Abrams and the Army of His Times* (New York: Simon and Schuster, 1992), p. 245.
60. Andrew Bacevich, pp. 6–7. For a thorough assessment of OPMS 71, see William Hauser, "The Peacetime Army: Retrospect and Prospect," in *The United States Army in Peacetime*, Robin Higham and Carol Brandt, eds. (Manhattan, KS: Military Affairs/Aerospace Publishing, 1975), p. 217. Also see, David McCormick, *The Downsized Warrior: America's Army in Transition* (New York: New York University Press, 1998).
61. Army Research Institute, *Newsletter*, vol. 9 (Alexandria, VA: The Army Research Institute, June 1992). The effect of "climate" on trust, cohesion, and unit effectiveness has been validated in many studies. The Army has considerable experience with a variety of surveys beginning with the *Study of Professionalism*. It is the use of data, not the gathering of it, that has been the primary flaw in the Army's survey efforts.
62. Donald Vandergriff, "Revolution in Human Affairs," unpublished briefing, retrieved from http://www.d-n-i.net/vandergriff/rha/index.htm, April 26, 2005.
63. Walter F. Ulmer, Jr., "Military Leadership into the 21st Century: Another Bridge Too Far?" in *Parameters* (Carlisle, PA: U.S. Army War College, Spring 1998), pp. 4–25.
64. Allan R. Millett, *The General: Robert L. Bullard and Officership in the United States Army 1881–1925* (Westport, CT: Greenwood, 1975), pp. 208–210. Millett talks about LTC Robert L. Bullard's perceptions of the influence of the emerging culture of management science: "Bullard thought that modern society had made leadership especially difficult because it produced a 'common lack of manly honor and self-respect' among soldiers by its impersonality and its materialistic value system." Millett goes on quoting Bullard, "The straining life of highly organized society has undoubtedly made men more nervous, more hysterical and less able to face danger, suffering and death . . . the frequent necessity of defense of self and rights, have made them more than ever loth [sic] to risk their lives in war and battle." Also, Russell F. Weigley, *History of the United States Army* (Bloomington, IN: Indiana University Press, 1984), pp. 598–600.
65. Retrieved from http://www.army.mil/thewayahead/quality.html.
66. The Honorable Les Brownlee and General Peter Schoomaker, "Serving a Nation at War: A Campaign Quality Army with Joint and Expeditionary Capabilities," *Parameters* (Carlisle, PA: U.S. Army War College, Summer 2004), retrieved from http://findarticles.com/p/articles/mi_m0IBR/is_2_34/ai_n6082900.
67. General Casey became the 36th chief of staff, United States Army, on April 10th, 2007.

68. For more discussion on Modularity, go to http://www.fas.org/sgp/crs/natsec/RL32476.pdf.

69. Donald E. Vandergriff, *Path to Victory: America's Army and the Revolution in Human Affairs* (Novato, CA: Presidio Press, 2002), pp. 115–160; this chapter discusses, in detail, why unit manning as a whole failed, but specifically COHORT.

70. Retrieved from http://www.army.mil/fcs/. For critical review of FCS, go to http://www.fas.org/sgp/crs/weapons/RL32888.pdf.

71. ARFOGEN can be found at http://www.army.mil/APS/06/maindocument/print/Text_addendum_E.html.

72. Details on growing the Army can be found at http://www.ausa.org/webpub/DeptHome.nsf/byid/JRAY-6ZBJDE.

CHAPTER 5

Recruiting Tomorrow's Soldiers and Leaders

> "Wars are fought and won by men, not by machines. The human dimension of war will be decisive in the battles and campaigns of the future just as it has been in the past."
>
> FM 100-5, Operations, May 1986[1]

Fourth Generation Warfare of the post-modern era demands sophisticated approaches to war-fighting and, more importantly, war-winning. Changes in the security dynamics of the post-modern international environment should be enough evidence for our senior leaders to convince those in Congress and academia to support the need to radically adjust recruiting, pre-commissioning education, and initial entry and advanced training from a system that has been accepted as an extracurricular or club activity to one that shares the respect of accepted academic disciplines. It is imperative to round out the educational experience of Army cadets and recruits with more academic rigor and innovative training, including a foundation in "how to think" and decision making, while remaining focused on the martial, tactical level of war-fighting across the spectrum of warfare. Fortunately, this is beginning to happen, but it must be carried out and continued as accepted evolution.[2]

It is first necessary to examine today's recruiting practices. Army leaders and soldiers (all are considered recruiters) are doing the best job they can given the market competition with society, an Industrial Age Army bureaucracy, and an unpopular war; but there are still some things that can be done, not just in recruiting. As discussed later in this chapter as well as in Chapters 6 and 7, the Army is already beginning serious changes in recruiting practices, training doctrine, professional military education, and overall force structure. The Army must continue to evolve these institutions for the long-term benefits, not just for recruiters, but in an Army better prepared for full-spectrum conflict, be it Third or Fourth

Generation warfare. Before we examine what the Army is doing to man its legions, let's look at current challenges.

In January 2007, the Army was given the go-ahead to expand by 35,000 soldiers by 2013. The Army is confident it can grow by 7,000 additional recruits a year, providing it has the necessary variety and size of incentives in place. That said, the U.S. Marine Corps also is growing, and it will be a challenge to attract an increasing number of recruits from the same pool without lowering standards. The cost of incentives is huge—over $600 million in Fiscal Year 2006 and rising. However, the Army exceeded its target. Attracting a greater number of recruits in 2007 through 2012 will require more money. There is no doubt that the top priority of the government, Department of Defense, and the Army is to reach the targets to produce a fully manned Army at war, relieving the current strain, rather than reduce the targets to fit the purse.[3]

As well as requiring more soldiers, the Army also says it needs more officers (this is an allegation with which I do not agree, and will discuss later in the book). The current total of 40,300 junior officers needs to grow by 3,000 in the next year, with the current attrition rate, at the higher end of the historical norm—but by no means exceptional—at 7.9 percent and in a testing operational environment involving multiple tours of Iraq. This will be challenging.[4]

The Army says that the conversion to the modular force discussed in Chapter 4 requires an additional 500 captains and majors. Also, repetitive operational tours of 15 months with a tour interval of only one year have caused the junior officer attrition rate to rise. Though still below the historical norm, the figure is seen as too high by human resource policy analysts in the Department of Army G1 (personnel).[5]

With the United States Military Academy at West Point restricted by law to 4,000 cadets and a Reserve Officers Training Corps (ROTC) program of two to four years, the only possible short-term solution is to increase commissions from the ranks. As a result, Officer Candidate School (OCS) increased its intake by a further 200 a year in June 2007. This is a short-term solution, while efforts and policies are directed to address medium- and long-term solutions.[6]

Focusing on the near term creates issues with quality in both enlisted and officer recruits within a culture that sees the mission or end state as the end all. With several recent examples of serious offenses being committed by some soldiers from lower educational categories, close attention is being paid to numbers of recruits from Category 4, the lowest educational category. There also is an increasing use of "waivers" for potential recruits with minor criminal convictions:

> The review, in its early stages, comes as the number of Army recruits needing waivers for bad behavior — such as trying drugs, stealing, carrying weapons on school grounds and fighting — rose from 15 percent in 2006 to 18 percent this year. And it reflects the services' growing use of criminal, health and other waivers to build their ranks.

> Overall, about three in every 10 recruits must get a waiver, according to Pentagon statistics obtained by AP, and about two-thirds of those approved in recent years have been for criminal behavior.[7]

The trend to lower standards is appearing on the officer side as well. There have been adjustments to give cadets more time to make the Army's height and weight standards as command pressure on ROTC programs grows to commission more cadets. At the same time, it is more difficult for cadre to eliminate poor performers from their programs—the counseling paper trail has to be overwhelming, while command approval for dismissal has grown tougher to obtain. This trend slowly vibrates throughout other Army courses, such as the cadre at the Basic Officer Leader Course (BOLC) II, which cannot eliminate poor lieutenants from the program, but must pass them on to units to take care of the problem.[8]

Finally, the way war is waged, as alluded to in the Introduction and Chapter 1, is demanding more than ever before out of privates as well as generals. Key senior officers—notably General Peter Schoomaker, former Army Chief of Staff—recognized that the U.S. Army was culturally in the Industrial Age and was unsuited to the complex demands of future warfare, and will need new operational ethos, as well as new structures and technologies. General Schoomaker worked hard to get this message through to the troops, albeit he clearly understood that this would be hard going, because it cuts across some deeply ingrained traditions, not only of the Army, but of U.S. society as a whole.

Recruiting Today

> "This recruiting problem is not just an Army problem, this is America's problem. And what we have to really do is talk about service to this nation – and a sense of duty to this nation."
>
> General Richard Cody, Army Vice Chief of Staff[9]

U.S. Army recruiting is executed by the U.S. Army Recruiting Command (USAREC) based at Fort Knox, KY. It is divided into five brigades consisting of 41 battalions controlling 1,684 recruiting stations. USAREC falls under U.S. Army Accessions command, which, in turn, falls under Army Training and Doctrine Command located at Fort Monroe, VA. U.S. Army recruiting policy and practices are formulated by the assistant secretary of the Army for Manpower and Reserve Affairs and the deputy chief of staff G1 in the Pentagon in Washington, DC.[10]

The recruiting environment is dominated by the All Volunteer Force (AVF) at war, negative media coverage, and the greater role of "influencers" and "gatekeepers." Interestingly, the aspiration of a large number of high school students to attend college is already affecting recruiting. In addition their parents and influencers, like coaches and teachers, also have a large input in the decision whether or not to enlist. As a result, the U.S. Army is targeting these students with a revolutionary program under its new Army Incentive Model (AIM):

> The Army has been enlisting youths for decades by promising them money for college. Starting in January, it will try out a different sort of pitch in selected cities: offering up to $40,000 toward the purchase of a home or the creation of a business. The new recruitment program, dubbed the "Army Advantage Fund," is meant to show parents and other adult "influencers" that Army service offers tangible benefits to young Americans. As the Iraq war continues, the Army is struggling to recruit enough new soldiers – and such influencers are less and less likely to recommend military service to youths. "If you want to get a soldier, you have to go through mom, and moms want to know what kind of future their children will have when they leave the Army," said Lt. Col. Jeff Sterling, the program's architect. "This is meant to answer that question in a tangible, concrete way."[11]

There is good reason for the Army to apply unconventional approaches to recruiting. Of a total market of 30.8 million 17–24 year olds, only 1.9 million are qualified to join the U.S. Army. The U.S. Army has 1,684 recruiting stations, 6,279 regular Army recruiters, and 1,774 USAR recruiters.[12] These focus on those regions and states that traditionally have recruited well, such as Texas and most of the southeastern states. They do not put much effort into recruiting soldiers from those areas that historically have not provided recruits. The recruiting targets for Fiscal Year 2007 (which ended September 2007) were

- Active component's (AC) annual target was 80,000, with a provisional target of an additional 3,000 soldiers following the announced intention to increase the Army.
- The target for the USAR was 26,500, with another 3,000 soldiers leaving the AC recruited to the USAR.
- The annual target for the National Guard (ARNG) was 65,100; in addition, 4,900 soldiers leaving the AC were to be recruited by the ARNG.[13]

The Army uses several methods to recruit soldiers. AC military recruiters do two-year tours. Each recruiter has a monthly target to meet—it is his or her "mission," and they must meet it to be successful. In addition to a financial incentive of $2,000 to soldiers who elect to become recruiters, the U.S. Army has put a good deal of effort into changing the profile of the position to remove the impression that it is a dead-end job. It is now regarded as a welcome break from operations—though recruiters work 12- to 16-hour days and sometimes 7 days a week.[14]

The Army also uses 400 civilian recruiters from contracted firms consisting of individuals attached to military recruiting offices. This program is a three-year trial in which two civilian firms recruit Army personnel. Most of the new civilian recruiters are locally recruited, retired Army, or members of the reserve. It is claimed that this initiative requires as many as 20 percent recruiters, who are cheaper than their military counterparts. This initiative is still in its infancy, but is likely to grow given the strain on the Army AC force as the war continues. As of this writing, civilian recruiters have signed up 15,000 recruits in four years.[15]

Civilian recruiting firms are paid $5,700 per recruit, with up to $1,000 bonus being paid to the recruiter. In 2006, one recruiter made $10,000 in a month. After initially disappointing results and poorer quality recruits, the standards and rules have now improved, resulting in a probable expansion of this concept.[16]

USAREC has also embraced the Internet. The U.S. Army's virtual "ibot" recruiter was launched in October 2006, and is on course to become the most successful recruiting tool in 2007. The "ibot" is called "Sgt STAR"—standing for Strong, Trained And Ready"—and is a virtual soldier who answers questions at GoArmy.com. Sgt STAR gives short answers to potential recruits who ask questions about the Army, but also pulls up an associated page with more information.[17]

Of course the Army also uses advertising. The U.S. Army has a large budget for advertising. The latest theme is "Army Strong," emphasizing individual, team, and national strength through individual service:

> The new slogan, developed in numerous tests with focus groups and interviews with soldiers, is meant to convey the idea that if you join the Army you will gain physical and emotional strength, as well as strength of character and purpose.[18]

In addition to advertising, the Army is also using mentorship. Serving and retired soldiers are encouraged to mentor potential recruits whom they know. Alternatively, they can register as a mentor on a database and will be allocated a potential recruit to guide and answer questions. This is not as forceful as it initially appears, is very well respected, and does have huge benefits for retaining interest in the Army. In addition, the mentor may well qualify for a referral bonus if he or she is a former soldier. These mentorships sometimes continue well into a soldier's career.[19]

By law, Army recruiters must be permitted access into schools and colleges. A few schools object to this and have challenged it in the courts. Generally, these are the type of schools and colleges from which few students have historically been recruited, such as Ivy League schools, and therefore the DoD has not pressed the issue. In addition, recruiters are permitted details of the students, allowing them to cold-call home numbers. This has proved more controversial, but the majority of schools comply, and it is an effective tool.[20]

One last thing that recruiters have to keep in mind, which is always asked by potential recruits, is "how long do I have to serve?" The standard term of service is four years with the AC and an additional four with the RC. In some specialist areas, a recruit can adjust this service to three and five years or two and six years—high-demand linguists and doctors for example. It is easy to transfer from the AC to the RC and vice versa. No part-time service is permitted while in the AC. While the system is now flexible, the time of service is still a negative point with many potential recruits because of their own misinterpretation or wrong input from their "gatekeepers," but recruiters have many incentives to offset the determent of service obligation.[21]

Tools of Recruiters—Incentives

Studies conducted by the Army consistently show that bonuses and education incentives are cost-effective in offsetting recruiting challenges posed by a strong economy or changing enlistment trends. They are used to expand the market, to channel recruits into critical skills, to encourage specific terms of engagement, and to reward higher education. When applied efficiently, incentives have proved the difference between failure in Fiscal Year 2005 and outstanding success in Fiscal Year 2006. The opinion of the DoD is that while fighting in Iraq and Afghanistan continues, and the Army expands, it cannot afford to fail in this area.

Four important factors should be understood in order to establish the effectiveness of U.S. Army incentives:

- A soldier can purchase a decent, basic three-bedroom house for an average of $150,000 in a garrison town.
- Healthcare is provided free for U.S. servicemen, but is not universally free for civilians. Though there are schemes for the poor, most people have to pay significant amounts of their income into healthcare.
- Further and continuous education is part of the working culture.
- Though the pension after 20 years of service is excellent, there is no lump sum payment on retirement.[22]

A wide range of financial recruiting incentives are principally targeted at the AC. Recruiting bonuses have increased from an average of $6,354 per entitled soldier in 2003 to $11,426 per entitled soldier in 2006. In 2003, 28,359 recruits out of a total of 72,000 were eligible for a recruiting bonus. In 2006, 44,019 out of 80,000 were eligible.[23]

An innovation was a referral bonus introduced in 2006. It is open to anyone, including retirees who refer a person to the Army, and they are paid $2,000 when the soldier completes 10 weeks of basic training. Military members of USAREC and close family members are not eligible for this bonus. However, USAREC is considering dropping the restriction on close family members.[24]

Every three months, a list is published in the military and recruiting press of trades/military specialties that attract a recruiting bonus. The financial incentives target the shortages. As soon as the trade meets its target, the incentive is diverted elsewhere. Therefore, in theory, a particular trade might only attract a financial incentive for three months. These incentives are widespread and targeted in detail. A recruit volunteering for a four-year engagement as an infantryman, a Multiple Launch Rocket System crewmember, or a satellite communications system operator will currently receive an incentive of $20,000, half which is payable on completion of training, with the remainder being paid in installments. Those who have served before will receive an additional bonus. The maximum combined bonus per individual is $40,000.[25]

Education is another important incentive. Education grants to assist further education are available to all soldiers. This is known as the Army College Fund, which is part of the Montgomery G.I. bill. Soldiers can obtain a grant of up to $72,000 for college after six years' service in the AC, as long as they match the grant with their own contributions over the term of service. This grant could soon be transferable to spouses.[26]

As mentioned at the beginning of this chapter, the Army is being innovative using incentives such as the Army Advantage Fund (AAF). The AAF is a financial recruiting incentive of $45,000 that targets buying a house and starting a small business. The aim is to satisfy a growing desire in the United States for home ownership, and to reinforce the potential for soldiers to start a small business as a second career. This plan will first be offered to recruits in 2008. In return for a four-year enlistment, a recruit will be offered up to $45,000 tax free to use as he or she pleases. It is thought that this will be available at the end of the four-year term, but this has not been finalized. A pilot plan will be introduced in early 2008, in which $30,000 will be offered to 500 recruits. The Army continues to use various forms of bonuses to lure potential recruits.[27]

Other bonuses include $3,000 for desired civilian skills, $8,000 for recruits who opt for OCS (commissioning from the ranks), and an additional $400 a month for a posting to a base that is unpopular and short of soldiers. Rather than opt for a trade that attracts an additional bonus, recruits can opt for a particular trade or location, or both, if they join.[28]

Other Factors to Consider

Taking into consideration the incentives required to enable a recruiter and the Army to be successful in manning its legions, the average cost of a recruit for AC has increased from $16,100 in 2003 to $21,500 in 2007 (and rising). This cost includes the pay of recruiting staff, advertising, recruiting operations, and bonuses. This large increase reflects a significant increase in the number of recruiters and bonuses. But, in the end, the U.S. Army continues to meet its recruiting targets in a challenging environment.[29]

The Army has established standards since the advent of the AVF that has given it a high-quality force. But, as the Iraq war continues, it has had to be flexible and modify its standards in order to fill its legions (more later on the impact of these changes). Some basic facts:

- The age limit for recruits was recently raised to 42 years old, resulting in an additional 900 recruits in 2006.
- Educational standards in the Army are well above the average for the United States, and high school graduation is the minimum standard for recruits. However, 2 percent of recruits (18 years old) are permitted to join without having successfully finished high school. The percentage has recently increased to 4 percent, but is unlikely to rise any further.

- The Army dismisses approximately 900 soldiers a year who fail to meet their weight and height standards. In an attempt to improve both retention and recruiting, the Army has conducted a complete review of the regulations. All recruits are measured for body fat. Those who fail the criteria were previously rejected. The new initiative allows those who fail this test to take the Assessment of Recruit Motivation and Strength, which tests strength and stamina. This resulted in 800 additional recruits in 2005. Trainees have six months to meet the Army's retention body-fat standards, which are stricter than enlistment standards.
- All U.S. Army soldiers are weighed, measured, and tested for fitness twice a year. Those who fail are placed on the "fat boy program."[30]

The Army also represents the U.S. society ethnically. The African-American population has always constituted a greater percentage of the Army than its 13 percent of the U.S. society as a whole. Recruiting of African-American soldiers has fallen dramatically, however, from 23 percent in 2000 to 13.5 percent in 2005. The Iraq war has become unpopular in the African-American community and there is a misconception that it is suffering high casualties—in percentage terms, it is suffering few casualties. In addition, fewer African-American males are graduating from high school at 18, but of those who do, more are attending college, thereby reducing the target pool of recruits.[31]

The female African-American population constitutes 37 percent of the females in the Army, reduced from 43 percent in 2000. However, the number of African-American female officers has risen to 23 percent, thought to be a gradual realization of the educational grants available. Women constitute 11 percent of the Army.

Since 9/11, there has been a large increase in the Hispanic recruiting, particularly in the infantry. Although Hispanics constitute 17 percent of the U.S. population, they form only 10 percent of the Army. However, the percentage has doubled in five years. It is thought that this is also due to the educational incentives. In the current Iraq conflict, casualties among Hispanics are high, primarily due to a large proportion opting for the U.S. Marines or Army infantry.[32]

Alongside recruiting, attrition and its causes must also be addressed. Attrition rates from training and within the first three years of service remain historically low. First-term engagement attrition continues to show an overall decrease. The 12-month loss rate for soldiers in the first 0–36 months of service decreased to 10.62 percent in May 2006, a reduction of 4.28 percent from May 2005. Had the overall first-term attrition rate remained at the rate reported in September 2005, there would be approximately 5,000 fewer soldiers in the U.S. Army today.[33]

As of the end of May 2006, the U.S. Army discharged 5,251 initial entry trainees from a 12-month training input population of 69,148, a 7.59 percent training base attrition rate. This rate represents a decrease of 10.53 percentage points (58%) from the 18.12 percent rate reported for May 2005. The Army continues to face challenges, so it must continue to encourage and reward innovation. To ease the burden on recruiting, the Army is also looking at many ways to retain soldiers.[34]

Retaining Good Soldiers

One of the priorities of any effective organization, military or civilian, has to be the retention of quality people. The U.S. Army is at war, and the ability to retain young, combat experienced soldiers is one of its top priorities. There are themes running through today's retention campaign:

- Maintain the AVF. To maintain such a force costs a good deal of money to reinforce "Terms and Conditions of Service."
- Look after the soldiers' families.
- Continuing to serve the nation while it is under threat requires self-sacrifice. We ask solders to "do your bit." In return, the nation will look after the soldier, both serving and retired.

Retention was an outstanding success in FY 2006. Every soldier and officer signs on for a short tour of service of between two and five years, and must re-enlist or, for officers, continue to get promoted at the end of that term or leave the AC. Therefore, accurate retention targets can be set by headquarters, with commanders at every level held accountable for the success or failure in reaching the target. This retention target is an annual stated "mission" (as with recruiters) for every commander, sometimes down to company level.

For example, retention figures for FY 2006 were

- The AC target was 64,200, while 67,300 re-enlisted, 105 percent of the target.
- The USAR target was 17,700, while 18,200 re-enlisted, 103 percent of the target.
- The ARNG target was 34,900, while 41,100 re-enlisted, 118 percent of the target.

The retention effort in FY 2006 was targeted on those soldiers with 6–10 years of service. This was considered to be the most likely group of individuals to leave in the current environment. Factors that affect retention are financial retention benefits, promotion opportunities, duration and frequency of operational tours, entitlements and quality of life, and the state of the economy. Taking these factors into consideration, the retention campaign for 2006 was repeated in FY 2007.

The FY 2007 retention policy is guided by

- Continued personal involvement of commanders and senior leaders in their retention programs and well-being of soldiers and families.
- Continued monitoring of the effects of multiple operational deployments upon retention.
- Closely monitoring the effectiveness of financial incentives.
- Monitoring and encouraging AC soldiers who transfer to the RC.

Most importantly, the Army has realized that its "current Army incentive program is too rigid, and not flexible enough to appeal to the new Millennium Generation."[35] The Army implemented a new AIM in the summer of 2007. The AIM is flexible in that it allows soldiers taking incentives to switch them to other incentives as they mature in the Army. Soldiers are not "bonded" to their earlier choices made during their enlistment or re-enlistment.

Again, as with recruiting, the centerpiece of retention is financial, but closely followed by intangibles based on what keeps soldiers in the Army. Roughly 318 million dollars were spent on financial bonuses related to re-enlistment while deployed on operations. The total expenditure on financial re-enlistment bonuses was $725 million. These factors are broken down further subsequently, among various innovative programs.

In 2003, it was possible for all soldiers who re-enlisted on operations to receive the bonus tax-free. This has been effective. Soldiers are able to re-enlist at the three-year point of a four-year engagement, and therefore time their move to coincide with operations. Alternatively, some soldiers volunteer for operations as they approach the end of their enlistment. A total of $318 million was spent on operational bonuses for re-enlistments in 2006.[36]

In 2003, the average re-enlistment bonus was $8,878 for each of the 10,035 soldiers signing for a further four-year term. This was from a pool of about 60,000 soldiers who re-enlisted. In 2006, the average re-enlistment bonus was $14,085 for 39,610 soldiers out of 64,200 who re-enlisted. This was a significant increase.

There are also bonuses for other actions beyond re-enlisting for time periods. For example, there are financial incentives for soldiers of the AC who transfer job specialties. Some of these are up to $40,000, in cases such as transferring into Special Forces. There is also a critical skills retention bonus to retain specialist senior non-commissioned officers (SNCOs) for up to five years beyond their 20-year pension point. This is paid predominantly to Special Operations Forces (SOF) SNCOs. To date, more than 350 soldiers on operations have received this bonus at an average payment of $72,000. An SOF SNCO can receive a maximum bonus of $150,000 for an additional six years' service.[37]

Another program is the Blue to Green, under which Air Force and Navy personnel approved for transfer can move to the Army without changing rank or discipline. Those who sign on for three years can receive a maximum bonus of $15,000. This program aims to capture personnel leaving the Navy or Air Force as a result of the drawdown of those services. The Navy is reducing by 60,000 and the Air Force by 40,000.[38]

For the first time in the history of the Army, officers were also offered financial incentives. Captains have five incentives from which to choose, with a limit of one option per soldier. Each option has to be "paid for" with a service obligation, for which the period of time varies, depending on the incentive:

- Cash (Critical Skills Retention Bonus): There are three tiers, depending on branch—$25,000, $30,000, and $35,000. The $35,000 bonus is limited to captains commissioned in military intelligence, infantry, field artillery, aviation, and the transportation corps. The service obligation for any bonus is three years.
- Graduate school: The Army will pay for an officer to earn a master's degree. The service obligation is three years for every one year of graduate school.

- Ranger school (requires a 12-month service obligation) and the Defense Language Institute (requires a three-day service obligation for every day of school).
- Switching career branches or functional areas: The service obligation is three years.
- Extending for assignment to installation of choice: Officers who take this option must have completed 12 months at their current permanent duty stations. If approved, officers will be guaranteed stabilization for at least two years. The service obligation is two years.[39]

The Army also offers non-financial incentives to both enlisted and officers. In addition to the financial incentives available to soldiers who wish to swap military specialties, a soldier who re-enlists is offered the choice of transferring to any arm of his choice. If this incentive is taken, then the soldier generally loses any other incentive. Soldiers are also offered choice of location as a re-enlistment incentive but, as in the previous case, by taking this option, the soldier generally relinquishes any other incentive available.

One of the other big incentives is an expansion in the area of further education. For officers, it guarantees a masters degree at a civilian university for signing up for an additional three to four years. As a result, some West Point cadets are signing on for eight years. In a country in which higher education qualifications are extremely valuable, this is an attractive incentive.

For all soldiers, a new element of the re-enlistment package allows soldiers to transfer educational benefits to their spouse. The spouse has been identified as a key element of the decision of a soldier to re-enlist. The new scheme targets soldiers who have served between 6 and 14 years. As an example, a soldier who wishes to re-enlist and transfer $15,000 of benefits to their spouse would have their re-enlistment bonus reduced from $25,000 to $10,000. The scheme will soon be expanded to include dependents—a very bold and imaginative move.[40]

Finally, the Army has made it easier to transfer from the RC to the AC and vice versa. This also encourages retention in the wider Army. The ARNG and USAR are now fully committed to supporting operations in Iraq and Afghanistan.[41]

Despite all these incentives, many of which are innovative and bold, the Army (as well as all the services) had to resort to using what is called a stop loss. The stop loss policy is the involuntary retention of some troops to remain in uniform beyond their expected or agreed term of service. It was created by Congress after the Vietnam War. Army recruits sign the enlistment contract on joining, which states that in event of war, the Army can keep them in uniform up to six months after the war ends.[42]

The current policy only affects those soldiers deployed in support of operations in Iraq and Afghanistan. Right now, about 10,700 soldiers are affected by this scheme; about 8,000 of them are AC. The Army uses the policy to retain key skills and enhance unit cohesion and experience on operations. With long operational deployments of up to 16 months, this policy is extremely unpopular with those it affects, and has been challenged both informally and formally, without success.[43]

The secretary of defense regards stop loss as an unnecessarily blunt tool with which to retain manpower and considers that better planning could achieve the same effect. However, the projected increase of the Army by 35,000, to 547,000 over the next five years is likely to seriously disrupt this plan, and it is likely that the policy will remain in force for the Army for the foreseeable future.

Despite the continuing major success in recruiting and retention, there is a concern that "retention fatigue" will begin to appear in 2008. Many soldiers who joined in 2001 and 2002 are approaching their second re-enlistment at the seven- or eight-year point. A high percentage of soldiers are on their third or fourth year-long operational tour. Despite the likelihood of increased and targeted incentives, it is expected that many experienced soldiers will leave in 2008. Interestingly, some believe this will accelerate as the United States begins to withdraw from Iraq, with soldiers feeling that they've "done their bit."[44]

Leading Tomorrow's Soldiers

The global research firm RoperASW asked Generation Y how they felt about their education and what they look for in leadership. They have seen trust rise to the top of the student and employee priority list. Supplemental studies of Randstad (a temporary services company) last fall found a further shift in the trust paradigm. Students and employees in droves now wonder not if the teacher or boss is *capable*, but if he or she is *honest*.[45]

Harvard Business Review points out further evidence that Generation Y expects trust and wants to be challenged. Skepticism puts a damper on innovation. Trust is even more critical in the future Army because of the strong tie trust has to innovation. The Army will depend on innovation for its future. But research shows that if employees don't trust, they don't innovate. Another reality of the times is that Generation Y closely associates loyalty with honesty.

Mistrust is pervasive, and it impacts the Army. Today's recruits and cadets see it in sports, the corporate world, and government. They see it coming out of the current war: Who is telling the truth about what? In the Army, they see double standards, from the magnified pressure to attain immediate results that are a holdover from the obsession with making mission, as people not worthy of promotion or selection move up anyway in order to meet numbers. The message: the end justifies the means, and today's youth don't like it or will not tolerate it as Boomers did.

My studies of effective armies and units have shown that those who balance competitive zeal with a passion for trust and reputation are those that win. What those armies and units understand is that the train must move on two tracks—competition *and* collaboration. Innovation is tied to collaboration. And collaboration is tied to trust. You can't achieve that by taking moral shortcuts. Without trust, the train is going to derail.

The principles of leadership—set the standard, enforce standards, lead by example, while caring for your people—remain constant. But there are particulars that will help recruiters and leaders deal with the Y Generation What these new soldiers bring to the table will assist the Army in moving into the Information Age.

When leading and developing future soldiers, check your assumptions. It's counterproductive—and usually misguided—to project your values on them. Talk about the "why" behind the Army's value system. Why is this important in the stress of combat?

Talk, don't attack their values. Allowing recruits' and cadets' true selves to shine within the need to impart discipline is hard, but valuable. Cadre and senior leaders must lead by example and show the Generation-Y cadet why the Army values are important.

Challenge them. Generation Yers love to learn—in fact, many experts tag them the most education-oriented generation in history. Don't assume that they have to be introduced to complexity through the crawl-walk-run method originated by Frederick Taylor more than a hundred years ago.

Mentor them. Feedback and support are essentials for Generation Y. These are not sink or swim Boomers. Generation Yers are insisting, "Get me up to speed now, so that I can add value." And they are grateful for all hands up they receive—as long as they aren't condescending. "Cadets want to know how to improve," says one cadet. But the strongest motivating technique may be to give cadet Yers room to do things creatively, to try and to fail. Adds another cadet: "The best cadre teachers give directions, step back, give directions, step back."

Twenty-two-year-old Second Lieutenant Jeff Fannelli praised the informal mentoring he received from cadre during his time in ROTC. "I felt like a professional peer," says Fannelli of one cadre. "He treated me like someone who was knowledgeable and he showed me respect even though I was a cadet." And the relationship didn't stop there. Adds Fannelli: "He went out of his way to show he cared about me as a person, even after I graduated."

Make it valid. Yers need to know how they will be affected by a decision or policy before swallowing it whole hog. For example, rather than simply telling cadets they have to do pre-combat inspections (PCIs), explain how such efforts benefit them directly: It's also important that Yers care about the mission of the Army. But what is going to be hard to swallow is that many cadre have to explain the "why" behind the way things are done in the Army. I must add the caveat that, in a learning environment, it is essential to explain why we are doing this. Says another cadet: "It's hard to stay motivated when you don't care about what's going on at your battalion."

The Cost of Dumbing Down?

Is the Army lowering standards, as many in the media claim, or is the Army doing the best it can given the limited parameters of today's environment, particularly in the case of an AVF within a society which is asked nothing more by its strategic leaders than to spend more, as pointed out in a recent security report, "The U.S. Military: Under Strain and at Risk":

> Although President Bush has sent the U.S. military to war in Iraq and to fight terrorism around the globe, he has failed to mobilize the American people for

> either cause. There has been no JFK-like "ask not what your country can do for you, but what you can do for your country" speech – no call to national service. This failure of leadership has only widened the gap between U.S. military personnel and the American people they are risking their lives to protect and contributes to the recruiting crisis.[46]

While the Army is doing all it can do to encourage volunteers of the quality it needs to face the evolving face of war, a quality which needs to be higher than ever before given the type of demands war is putting on their shoulders,

> Our recruiting campaign is bringing into the Army thousands of new soldiers (perhaps 10% of the annual input) who should not be in uniform. (Criminal records, drug use, moral waivers, non-high school graduates, pregnant from Basic Training and therefore non-deployable, lowest mental category, etc.)[47]

Not only is this seen from the top down, but also there is a lot of evidence from the ground level of an increased impact from lower standards and the strain it has on the Army, both on the enlisted and officer sides:

> The Army is between a rock and a hard place – they've asked for something they are not going to be able to achieve. They aren't going to make the 547K growth without lowering standards for initial recruits even further, and I have no idea how to fix the Officer problem other than direct commissioning NCO's. The OPTEMPO is breaking the officer and NCO corps, despite what has been said. I know Majors in the AC that have done 6 deployments since they were commissioned in 1996 – Bosnia, Kosovo, Afghanistan X2, Iraq X2. I know Guard officers who have done three deployments in the same time period. If I was the Army leadership, I would start warning Congress today that there is going to be another massive bill on the table – in one form or another – to not only keep officers in, but to grow the officer corps.[48]

The Department of Defense has countered the growing voices that the Army and the entire military are lowering standards in order to meet their recruiting goals. In a four-page paper subtitled, "Myths versus Facts," the Defense Department argues that "nearly two-thirds" of all military recruits "are drawn from the top half of America in math and verbal aptitudes – strong determinants of training success and job performance."[49]

But military sociologist Dr. Charlie Moskos counters DoD claims that standards are not being lowered.[50] Based on his continual visits to units and many contacts, Dr. Moskos says that "the more dropouts [who enlist], the more discipline problems" the Army is likely to have. These days, he says, most new recruits are likely to be "whites from small-town America," generally less educated and with fewer job options. "They're not the suburban kids [in uniform]. They're from the exurbs," Moskos said. "We're not recruiting people from the better-educated segments of society."[51]

What is the impact on an Army that relies less on its educated portion of society? Given today's Fourth Generation environment, the impact can have strategic implications. However, it should also be noted that a majority of those that are serving represent some of the finest in our society, and the fact that a larger portion of people not meeting past standards are being brought into service does not or should not take away from the superb job being carried out by most soldiers. One must remember an old command adage I heard over and over again when I was in company command for 36 months: "Most of your time will be spent dealing with the smallest number of your soldiers." The same applies today. It will be up to Army leaders at all levels to maintain standards even while not having all the people they need. "It is always better to have no officer than a bad officer," as the Germans would say when justifying their high standards in officer accessions. In today's 4GW world, it applies to all ranks.

Revolutionary Changes at Evolutionary Pace

History has shown that readiness and reaction to emergencies were met only with conscription, but at the same time, the volunteer military has produced an excellent force. The arguments for both have already been touched upon, but why just have one or another? Why not take parts of both, to maintain the ability of the military, particularly the Army, to accomplish its global tasks?

According to Mr. James Larson of U.S. Army Accessions Command, "There are short-term incentives the Army can work on that deal with the tactical arena of manning." Mr Larson went on to state, "In the future, advances in automation tools that highlight specific available unit positions combined with screening tools that identify accurate matches for open positions (closely matched with soldier quality feedback mechanisms from military units) will allow the Army to access an AVF capable of regular missions at the minimum market cost."[52]

Another set of screening tools and mechanisms will be created to support all unusual military surges that exceed the capabilities of the AVF. In order to accommodate larger active military forces during conflicts and larger reserve forces during peacetime, the three components of the Army—Active, Reserve, and Guard—will have to develop a seamless transition program that allows soldiers to move among components without penalties.

Mr. Larson pointed out that

> A future set of accurate and predictive assessment screening tools would enable the force to determine whether an individual meets the minimum requirement for a specific position. Well-developed feedback from units (with a warranty replacement system to resolve the occasional problem of soldiers incapable of performing a specific mission) will enable the accessions missions to fine tune the requirements for each position based on a variety of physical, psychological, linguistic, social, and mental measurements to offer each interviewee a better look at jobs for which they can contract immediately.[53]

With accurate and low-cost screening mechanisms and tools, a potential soldier who fails to meet the requirements for employment in a job he wants can be retested at low cost after the potential soldier makes changes (physical or otherwise). Automated equipment linking potential soldiers to accurate information on the availability of positions will allow a variety of social and official networks to fill positions with more soldiers, more accurately matched (in terms of skills and abilities), with a less extensive network of official recruiters and more individuals operating for other motivations to fill the force.[54]

To take advantage of improvements in technology, each seat that the Army is hiring to fill in a unit (Modified Table of Equipment or Table of Distribution and Allowances [TDA])[55] should be individually compensated and positions given incentives depending on the importance of the slot to the total Army mission. At various intervals, performance bonuses can be provided to soldiers depending on their ability to meet job requirements and their perceived ability to perform at higher levels of responsibility. To leverage a more educated work force, "the Army can recruit based on civilian-acquired skills, especially in technical areas, where soldiers can be recruited at higher ranks and given credit for their civilian experiences." Positions throughout the Army can be bid to all soldiers from the active Army, Reserve, and Guard who have met the requirements of a prerequisite position, allowing for a variety of negotiations for pay and incentives.[56]

In order to accommodate an ever-changing society, the Army must tailor its incentive packages to a "cafeteria style" approach. That is, there may be benefit derived from offering a combination of monetary and nonmonetary incentives and allowing applicants to choose from bonuses, health care, sabbaticals, and telecommuting opportunities that best fit their situation. In addition, it may even be necessary to allow some of these incentives to be passed on to spouses and/or children.

Currently the Army allows prospects to contract for specific occupational specialties but be worked in whatever manner that the Army determines for up to eight years. The Army keeps a great deal of flexibility for itself in assigning its personnel. The cost paid for this flexibility (in terms of location of service, duration of service, pay for rank and years of service, no retirement until a full 20 years of service, and assignment flexibility) come at a cost when individuals negotiate for a position in the AVF.

To encourage future service member to accept a lower pay check than the labor market would set for jobs with similar risks and similar expected return on their individual's investment of time and effort, a variety of recruiting incentives and fringe benefits have been used. "Examples include enlistment bonuses for hard-to-fill jobs, educational opportunities both in and following service, specialty training unavailable to civilians, MOS training with dual military and civilian value, medical treatment, and a unique retirement program." A multi-billion-dollar DoD expense to recruit the workforce necessary to defend the nation can be reduced primarily by matching individuals best suited to specific defense work with the highest degree of accuracy.[57]

Regimental System and Depots

The Army can also change the way it recruits from the Industrial to the Information Age, still taking advantage of approaches validated through the ages, including regional pride and cohesion. A major change inside recruiting is decentralization. The diversity of the larger American culture does not eliminate the possibility of employing a regionally based regimental system that rotates and replaces units in and out of full-spectrum missions. The regional regimental depot system has been highly successful in the British, German, Canadian, and Israeli armies (which employ the Swiss reserve system focused more on smaller units—which are currently carrying the burden at the squad and platoon levels in the current evolution to war). The British Army has the most experience with the system, employed successfully in linear and nonlinear warfare.[58]

A regimental depot system addresses both the intangible and tangible issues regarding manning. First, it promotes unit cohesion at the institutional, organizational, and small unit levels. Unit cohesion gives units more agility and psychological strength in combat. Cohesive units also help significantly to solve many costly problems, including large amounts of money required to maintain training readiness to counter turbulence caused by numerous individual changes of station (PCS) moves. Most importantly, cohesive units will be prepared to conduct the multitude of complex operations delivered by the changing face of war—revisiting soldiers' baseline skills takes shorter periods, enabling units to deal with new and more complex unit tasks. Regional regimental depots supply units with incremental replacements while also supporting the Army Generational Force model discussed in earlier chapters.

The regional regimental depot will establish initial organizational bonds among the soldier, the regiment, and the region. It will welcome new recruits into the family ceremoniously, after being recruited by soldiers already from the region. Each regiment will have an administrative headquarters based at an Army post in the United States that supports AC and RC units from the same regiment (similar to the British Territorial Unit system), and it will draw most of its personnel from the geographic region. But, like major collegiate football today, recruiters from a regiment may recruit candidates from other regions as well, using innovations and incentives, as well as the reputation of the battalion's task forces.

The depot will also provide a home for soldiers by giving them a geographic identity on a scale sufficiently small to be significant to the individual and his or her family. It will reduce the impersonal approach of the current centralized army personnel system, making possible meaningful career planning inherent in a decentralized institution. Additionally, the depot regiment will have a training battalion responsible for basic and advanced individual training, and a headquarters and headquarters company to manage administrative tasks.[59]

There Is Also Room for Conscription (Limited)

There are also revolutionary ideas that the nation can implement that support the Army at the strategic level. While the volunteer force would continue to be maintained, it is now time to offer conscription under a new three-tiered approach pushed forward in 2001 and again in 2004 by Dr. Charlie Moskos—15-month enlistment—coupled with a separate proposal recommended by John Tillson of the Institute of Defense Analyses called "college cohort." Under these plans, the nation can use limited conscription but also move to recruit a cohort of "buddies" for only 18 months, and then return them ready trained as well as experienced to the U.S. labor market.[60]

Tillson's approach would also cater to groups of college-aged men and women who have already or are bonding. It would allow them to enlist as a small unit, such as a squad or even as a platoon for 15 months, and then return to what they were doing before their service. In return for the incentives given now for college and pay, the group would be kept together for their entire service obligation, and could also opt for the same duty obligation in the ARR, if they so chose as a group.

Dr. Moskos' 15-month enlistment approach includes

- A 15-month enlistment option with generous educational benefits.
- Counterbalancing the prevailing econometric approach with social-psychological theories in recruitment. For recruitment purposes, nothing would be more meaningful than one's friends joining the Army.
- Considering a cohort enlistment for certain colleges to serve in a specified peacekeeping mission.
- Emphasizing military service as a rewarding experience between undergraduate and graduate school.
- Linking all federal aid to college students to some form of national service. We have created a G.I. Bill without the G.I.

Dr. Moskos sums up the benefits of this proposal:

> Let us also keep in mind the long-term benefits for the country if military service became more common among privileged youth. We will have future leaders in civilian society with a rewarding military experience—and who will be future informal recruiters. This can only be to the advantage of the armed services and the nation.[61]

This approach to recruiting would be within a new three-tiered system of youth service in America, with 15-month terms of duty for all citizens aged 18 to 25. In this new-style draft, conscripts would have what all Americans now demand: choice. They could choose to serve in the military, in homeland security, or in a civilian national service program such as AmeriCorps (there's no reason women couldn't be drafted for the latter two categories). In return, draftees would get G.I. Bill–style college scholarships, with higher awards for those who accept more dangerous duty.[62]

The first tier of the system in the 15-month period would not have to be offered the relatively high wages and benefits that it takes to lure voluntary recruits (an increasing number of whom are married with families). This would leave more funds available to raise pay for Army specialties and professionals such as computer specialists, mid-level officers, and master sergeants. The Army is paying recruits a lot of money to lure them into service, while those who have committed themselves, the underpaid sergeants, watch in awe. In the draft era, the pay ratio between a master sergeant and a private was seven to one; today it is less than three to one.

Restoring something resembling a meritorious approach to pay is the best way to upgrade retention in hard-to-fill skills and leadership positions, while returning to a true meritocracy. The best way would be to require all young people to serve. One reason more young people don't serve now is the fear that while they're wearing the uniform, their peers will be out having fun and getting a leg up in their careers.

"If everyone were required to serve, no one would feel like a sucker. They would benefit from the experience"; Dr. Moskos said that surveys show that former draftees look back on their time in the service with fondness and pride. In addition, today's youth would serve in the last bastion of selfless service and would come to understand that there is something more important than one's self. It's possible, however, that the country won't have the need for every eligible young person to serve. What then? One answer is a lottery with no student deferments. (Under Selective Service rules established after Vietnam, college deferments are no longer allowed.)[63]

Conclusion: Strategic Leaders Lead by Example

No matter what ideas are proposed here or by the Army in order to recruit and retain the right people, strategic leaders must convince the public that it is in their best interest to serve the nation selfishly. This is something President Bush failed to do right after 9/11 attacks. The president had the moral high ground, so why did he not do it? Why did he not believe enough in his own strategy to ask his followers, the American people, to also demonstrate their resolve through self-sacrifice? As discussed in Chapter 3, this plea has become harder and harder, particularly when leaders at the strategic level are also putting self-interest ahead of selfless service, as a continual laundry list of scandals attest. Only those in the services, and particularly the Army and Marine cultures, appear to attempt to adhere to the time-honored principle of selfless service, and that is why they are doing a bigger service to our nation beyond simply fighting its wars.

Yet, as seen in this chapter and the next one, the Army is far from perfect. A lot of work, innovative ideas, and bold leadership remain to be done for the Army to be effective dealing with the changing face of war.

Notes

The author is indebted to Mr. James Larson of U.S. Army Accessions Command for the background information in this chapter.

1. Department of the Army, *Field Manual 100-5 Operations* (Fort Monroe, VA: U.S. Army Training and Doctrine Command, May 1986), p. 3.

2. Morris Janowitz, "Civic Consciousness and Military Performance," in *The Political Education of Soldiers,* ed. Morris Janowitz and Stephen D. Westbrook, (Beverly Hills, CA: Sage, 1983), p. 76.

3. Greg Bruno, "West Point Wants to Expand Number of Students," at Record-online.com (February 2007), downloaded from http://www.recordonline.com/apps/pbcs.dll/article?AID=/20070201/NEWS/702010337.

4. "Expanding the US Army," The International Institute for Strategic Studies (February 2007), downloaded from http://www.iiss.org/ . . . /past-issues/volume-13—2007/copy-of-volume-13—issue-1/expanding-the-us-army.

5. Janet St. Laurent, "Force Structure: Capabilities and Cost of the Army Modular Force Remain Uncertain," Government Accounting Office (April 4, 2006), downloaded from http://www.gao.gov/new.items/d06548t.pdf.

6. Katherine Peters, "Looking for Good Leaders," Government Executive.com (August 1, 2007), downloaded from http://www.govexec.com/features/0807-01/0807-01na1.htm.

7. Steven Benen, "Military Forced to Lower Standards Once Again," *The Carpet Bagger Report* (November 7, 2007), downloaded from http://www.thecarpetbaggerreport.com/archives/13505.html.

8. This derives from discussions with 47 officer and NCO cadre over the past two years. Their biggest frustration is with not being able to eliminate poor-quality cadets and lieutenants, but having to pass them on to operational units as pressure from the chain of command has mounted to "make mission."

9. Scott Ritter, "The Military's Recruiting Problem," *Alter Net* (January 19, 2006), downloaded from http://www.alternet.org/bloggers/ritter/30941/.

10. Information on recruiting command downloaded from the U.S. Army Accessions Command website at http://www.usaac.army.mil/.

11. Yochi J. Dreazen, "Uncle Sam Wants You, but Ads Target Mom," *The Wall Street Journal* (November 29, 2007), p. B1.

12. Colonel Garcia, "State of the Youth Market," U.S. Army Accessions Command (Fort Monroe, VA: December 2006), slide 5, downloaded from http://www.cast.ilstu.edu/documents/MSC_FY08_Budget_Presentation.ppt.

13. Garcia, "State of the Youth Market," slides 5–7.

14. Interviews with Army recruiters.

15. "Civilian Recruiters," *About.com* (February 2002), downloaded from http://usmilitary.about.com/library/milinfo/milarticles/blcivrecruiters.htm.

16. Rod Powers, "Recruit Referral Bonus Program," *About.com* (March 24, 2007), downloaded from http://usmilitary.about.com/od/armyjoin/a/bonusrefer.htm.

17. Kelly Kennedy, "Virtual Recruiter Answers Questions," *Army Times* (January 13, 2007), downloaded from http://www.armytimes.com/news/2007/01/atsgtstar0108/.

18. Robert Strong, "Army Launching 'Army Strong' Campaign," Boston Globe.com (October 9, 2006), downloaded from http://www.boston.com/news/nation/washington/articles/2006/10/09/army_launching_army_strong_ad_campaign/.

19. Army Mentorship Resource Center, downloaded from http://www.armyg1.army.mil/HR/mentorship/docs/Army%20Mentorship%20Program%.

20. Charles Lane, "Court to Review Military Recruiting at Campuses," washingtonpost.com (May 3, 2005), downloaded from http://www.washingtonpost.com/wp-dyn/content/article/2005/05/02/AR2005050200616.html.

21. "Required Military Service Obligations," *U.S. Army Cadet Command*, downloaded from http://www.rotc.monroe.army.mil/helpdesk/enrollments-1/service%20obligations.htm.

22. Garcia, "State of the Youth Market," slides 13–17.

23. "Military Bonus Center," Military.com, downloaded from http://www.military.com/recruiting/bonus-center/.

24. Rod Powers, "Recruit Referral Bonus Program," About.com (March 24, 2007), downloaded from http://usmilitary.about.com/od/armyjoin/a/bonusrefer.htm.

25. "Army Enlistment and Reenlistment Bonuses," About.com, downloaded from http://usmilitary.about.com/od/armybonuses/Army_Enlistment_and_ReEnlistment_Bonuses.htm.

26. "Army College Fund," downloaded from https://www.hrc.army.mil/site/education/ACF.html.

27. Jerry Poink, "HQDA Army G-1 Incentives Branch," U.S. Army (February 26, 2007), downloaded from www.amc.army.mil/amc/pe/html/retention/PP_Files/AMC_SMART_Brief_1_Incentives_Briefing_FY07.ppt.

28. "Military Bonus Center," Military.com.

29. Lizette Alvarez, "Army Giving More Waivers on Recruiting," *New York Times* (February 14, 2007), downloaded from http://www.nytimes.com/2007/02/14/us/14military.html.

30. Updated Army demographics can be found at http://www.armyg1.army.mil/HR/demographics.asp.

31. Ibid.

32. Garcia, "State of the Youth Market," slide 7.

33. Ibid., slides 5–6.

34. Ibid., slides 5–6.

35. Lieutenant Colonel Kevin Beerman, U.S. Army, "Increasing Army Retention through Incentives," U.S. Army War College Research Project (Carlisle, PA: U.S. Army War College, March 15, 2006), p. 4.

36. Lolita C. Baldor, "Military Re-enlistment Bonuses Soar," *Oakland Tribune* (April 12, 2007), downloaded from http://findarticles.com/p/articles/mi_qn4176/is_20070412/ai_n19019173.

37. John W. Fountain, "New Incentive to Join Army: Direct Sign-up in Green Berets," *New York Times* (February 22, 2002), p. 15, downloaded from http://query.nytimes.com/gst/fullpage.html?res=9803EEDC113EF930A15751C0A9649C8B63&n=Top/Reference/Times%20Topics/People/F/Fountain,%20John%20W.

38. American Forces Press Service, "Operation Blue to Green," About.com (July 11, 2004), downloaded from http://usmilitary.about.com/od/armyjoin/a/bluetogreen.htm.

39. Lisa Burgess, "Army Captains Offered up to $35,000 to Stay," *Stars and Stripes* (September 19, 2007), downloaded from http://www.military.com/features/0,15240,149817,00.html.

40. "Military Bonus Center," Military.com.

41. *Army News Service*, "Transferring to Active Duty after Mobilization" (June 19, 2005), downloaded from http://usmilitary.about.com/od/guardandreserve/a/mobtransfer.htm.

42. Will Dunham, "US Army Uses Stop Loss to Force 50,000 Soldiers into Extended Duty . . . ," *The Huffington Post* (January 26, 2006), downloaded from http://www.huffingtonpost.com/2006/01/29/us-army-uses-stoploss-to_n_14667.html.

43. Dunham, "US Army Uses Stop Loss."

44. Beerman, "Increasing Army Retention through Incentives," pp. 4–5.

45. "RoperASW Interviews Y-Generation," downloaded from http://www.wmfe.org/site/DocServer/PBS_Kids-_Mktg.pdf?docID=441.

46. The National Security Advisory Group, "The U.S. Military: Under Strain and at Risk" (Washington, DC: January 2006), p. 4.

47. General Barry McCaffrey, "After Action Report—General Barry R McCaffrey USA (Ret.)" (West Point, NY: United States Military Academy, December 18, 2007), p. 10.

48. Major Andrew Dzenswinski, personal e-mails to author.

49. Department of Defense, "Who Is Volunteering for Today's Military: Myths versus Facts" (Washington, DC: Department of Defense, December 2006).

50. Bryan Bender, "Fewer High-Quality Army Recruits: As War Need Rises, Test Scores Drop," *Boston Globe* (June 1, 2007), downloaded from http://www.boston.com/news/nation/washington/articles/2007/06/01/fewer_high_quality_army_recruits/.

51. E-mail with Dr. Charlie Moskos, December 2001.

52. I am indebted to Mr. Jim Larson of U.S. Army Accessions Command for overwhelming me with an incredible amount of information on recruiting, developing, and training.

53. E-mail discussions with Mr. Jim Larson, December 2, 2007.

54. E-mail discussions with Mr. Jim Larson, November 29, 2007.

55. TDA units are organized to perform specific missions for which there are no appropriate TOEs and are discontinued as soon as their assigned missions have been accomplished. Unlike TOE units, TDA organizations are considered nondeployable, even when organized overseas, as their missions are normally tied to a geographic location. The personnel of TDA organizations can be military, civilian, or a combination of both. In some instances, provisional-type units have been organized under TDAs until suitable TOEs were established. Examples are some of the mobile army surgical hospitals (MASHs) and a ranger company organized in Korea during the Korean War. When the Army developed TOEs, the TDA organizations were discontinued.

56. E-mail with Mr. Larson, December 8, 2007.

57. E-mail with Mr. Larson, December 8, 2007.

58. LTC P. W. Faith and LTC D. I. Ross, *Application of the Regimental System to the United States Army's Combat Arm* (Fort Monroe, VA: Training and Doctrine Command [TRADOC], April 1980).

59. For more details on this proposal, see Donald E. Vandergriff, "Parallel Evolution," in *Path to Victory: America's Army and the Revolution in Human Affairs* (Novato, CA: Presidio Press, May 2002), pp. 209–223.

60. Discussions with Mr. John Tillson, February 2004.

61. Dr. Charlie Moskos, "The Citizen Soldier: The Ideal and Reserve Draft," unpublished paper (December 2004), pp. 2–3.

62. Discussions with Dr. Charlie Moskos via e-mail 12 December 2001.

63. Moskos, "Citizen Soldier," p. 4.

CHAPTER 6

Training (and Educating) Tomorrow's Soldiers and Leaders

> "The concept of Task, Conditions, and Standards may work for junior people in certain tasks. This concept, however, is becoming increasingly less relevant to higher order collective and leader training."
>
> "[The Army must] emphasize the importance of changing culture . . . change the frames of reference . . . move beyond Task, Conditions, Standards . . . reinforce development of a wartime mindset."
>
> "[We must] adapt the institutional Army . . . create a culture of innovation . . . counter resistance to change."
>
> General Peter Schoomaker,[1] Army chief of staff 2003–2007

For the first four years after 9/11, the Army remained focused on linear warfare, and only slowly changed from the bottom up after the shock of occupation following the fall of Baghdad. However, after a year of occupation in Iraq, the Army faced a new enemy, one that already had been evolving, and one that practiced Fourth Generation Warfare (4GW). Fortunately, there has been growing consensus that the new operational environment requires a new military culture and that this is going to be a "long war," in which traditional linear and attritionist approaches do not always work and the non-military lines of operation are centerpiece. Some important drags on changing the culture of the Army remain in place, four of which include

- Inconsistent level of professional military training and experience of company-level officers in the U.S. Army;
- Paucity of training opportunities at battalion and company level shaped by an obsession to retain an Industrial Age force structure and personnel system while trying to change everything else;

- Uneven application of genuine Mission Command with a still widespread culture of conformity and uncritical compliance;
- Simplistic, short-term, and culturally insensitive instincts of U.S. society at large.

The first bullet is focused on the officer corps, but enlisted training and development will also be addressed on more detail later in this chapter. There is no standardized entry test for U.S. Army commissioning. Ten to 15 percent of officer cadets come through the United States Military Academy at West Point, NY. Here, academic excellence takes priority over military proficiency and many of the places are allocated on the basis of Congressional patronage. Most of the rest of cadets (future officers) join through the Reserve Officers Training Corps (ROTC) located at 270 schools throughout the United States and its territories. A small, but growing, percentage comes through the 16-week Officer Candidate School at Fort Benning, GA. This course has been frequented more by former noncommissioned officers (NCOs) than by those who have met the minimum entrance standard with a degree and only basic training prior to attending, which is good for the Army if those former NCOs are not tied to the old way of doing things.[2]

The standards across these three entrance pathways are mixed, the newly established Basic Officer Leaders Course (BOLC) II at Forts Sill and Benning aims to level the playing field. As will be discussed later in this chapter, BOLC II is also employing a new learning and development approach for these new officers, but at six weeks, remains short of what many consider necessary for young officers to be prepared for 4GW.[3]

There are few training opportunities at battalion and company level. All individual and most squad-level training is directed by NCOs, to standards set by the sergeant majors and NCO academies, which have a chain of command almost independent of the officer corps. Young officers are principally responsible only for the moral component of the troops under their command. Their involvement in training is limited to authorizing exercise and safety standards, not for the conduct or the standard of the training itself, which are mainly the purview of the NCOs. This tends to produce repetitive, unimaginative, and risk-averse training, heavily tied to formatted training objectives, as well as further limiting opportunities for company officers to develop their own training and leadership skills. There is no tradition of unit or subunit organized Tactical Exercise without Troops or adventure training, both of which encourage junior leaders to develop self-confidence and independence of judgment.[4]

Finally, a top-heavy officer corps in which field grade officers' ratio to junior officers is one to one also discourages junior officer initiative. This is because the Army does not understand the critical impact the current out-of-date force structure and personnel management system have on the operational Army. These two institutions are in place today based on out-of-date assumptions, and they have become accepted as untouchable when speaking of military reform. Current organization and equipment are designed for Cold War–era "jousting matches" and conceptually simple conventional combat (besides enriching defense contractors). This greatly reduces the number of "boots on the ground" that current organizational models can yield.[5]

This all leads to the statement that there is an uneven application of Mission Command or Trust Tactics amid a widespread culture of conformity and uncritical compliance. This is, in part, driven by a bottom-up, staff-driven military decision-making process that encourages staffs to develop simple command choices for their commanders, rather than seeking the commander's intent and guidance up front. Finally, the American culture has a short-term focus, largely reflected in a Congress that has much more direct influence over military matters, particularly with technology, while having little understanding of the human dimension of combat or military effectiveness.

These are intractable issues, deeply rooted in U.S. Army culture, and there has been little appetite to address them until recently, as the war in Iraq and Afghanistan is forcing even the die-hard defenders of the status quo to question past methods. At present, the focus for change is at the level of doctrine, training, and force development; and in these areas, the signs of an institution in the throes of genuine cultural change are everywhere. Progress is inevitably patchy, with some branches and formations far ahead of others, as we will see later in this chapter, but this is largely dependent on the personalities and commitment of individual commanders.

The entire Army is now "talking the talk" and, increasingly, signs are appearing that it is "walking the walk." In some cases, the buy-in is simplistic, going little beyond mastering a new set of doctrine or "tactics, techniques, and procedures," but many now understand the deeper mindset changes that are required. Increasingly, U.S. Army officers will suggest that, despite the strains, current operations are making the Army better within its own confines of a 2GW structure and personnel system.

Change on the scale of organizational culture evolving inevitably takes time, but progress is being made, and what will be addressed here is how the Army used to train and develop its soldiers and why, followed by some new ideas that are being implemented that promise to better prepare the Army for 4GW. First, what does the Army want the future soldier to be prepared to do? Regardless of whether it is enlisted, NCOs, or officers, 4GW is flattening the baseline of what a soldier has to know to be successful. Specialization is going out the window, not back to the age of generalist, someone who knows a lot about little, but toward a new professional soldier, required to know many skills, taking parts of each to accomplish a multitude of missions. If this is the case, then is it not true that how those soldiers and their leaders are developed must change as well?

The Army today expects soldiers not only to be technically proficient in warfighting, but also capable of supervising civil affairs, providing humanitarian aid, and performing a range of activities relating to order and stability. As networked technologies flatten command structures, new doctrine and revised training regimes are going to be required in order to prepare individual soldiers to assume the greater responsibility of full-spectrum warfare that 4GW demands.

As a result of these trends, the U.S. Army must begin to foster a military culture that is aimed at preparing its junior officers and NCOs to become what has been described as "strategic corporals" or "strategic lieutenants." The term *strategic corporal* or *lieutenant* refers to the devolution of command responsibility to lower rank levels in an era of instant communications and pervasive media images. While there is no general agreement on the meaning of a strategic corporal (or lieutenant), for the purposes of this book and what Army leaders should look for, it is defined as follows:

> A strategic corporal or lieutenant is a Soldier that possesses technical mastery in the skill of arms while being aware that his judgment, decision-making and action can all have strategic and political consequences that can affect the outcome of a given mission and the reputation of his country.[6]

As the experience of the U.S. Army grows, it's getting more discriminating about its requirements. At the tactical level, they are increasingly interested in specific requirements—explosive ordnance disposal, search dogs, public order, close observation, negotiations. At the strategic level, they are genuinely receptive to the different insights and perspectives that allies bring. There is an increasing willingness to work together on doctrine and concepts (such as the initiative to align U.S. Army/Marine Corps/British Army Counter-Insurgency (COIN) doctrine) and to exchange experiences on structures, training, and forces generation. This is a far cry from the hubris of the recent past.

More History on Army Training (and Education)[7]

Commanding generals convene transition teams to explore critical questions about training and doctrine. The Army's Training and Doctrine Command (TRADOC) oversees training and soldier development.[8] The results are visions for the way forward into the future. Post–Desert Storm, senior Army leaders recognize the need for change and are not blind to the symptoms of inertia throughout the Army. They took action by creating special study boards, initiatives groups, and hand-picked protégés to help guide decisions on how to transform. However, also in common with other vision statements, ground observations regarding execution of the vision in the field reflect an effort that is still tied to rigid, time-consuming processes; questionable assumptions; insufficient resources (primarily the "right" people); not responsive or relevant to the needs of the operational Army; and too concerned with TRADOC internal protocols and priorities.[9]

To their great credit, senior Army leaders are fully aware of the perception. One of the most interesting pieces of evidence is the open dissemination of a TRADOC White Paper titled "Adapt or Die: The Imperative for a Culture of Innovation in the United States Army."[10] This paper was printed as an article in the Army magazine, a journal that is seen throughout the Army at all ranks. Finally, TRADOC made an audio tape about "Adapt or Die," expressing the need to change the culture. While the publication of the article is a start, there is no discussion of why it was published. In other words, there is no mention of the aspects of Army institutional behavior that indicate the Army does not have a culture of innovation. Also, the article "Adapt or Die" discusses the Army in aggregate and not specifics. What is not said, but becomes clear to the reader, is that while the operational Army has "transformed" in many ways since 2001, institutional Army organizational barriers to self-change are only now starting to be brought down.[11]

The Army has had disconnect between the intentions of its senior leaders and the execution to achieve those intentions by the organization itself. Reading the intent statements of TRADOC commanders, and indeed of any TRADOC subordinate organization, it is hard to find argument with their intent. However, intentions are only slowly evolving into practice in the eyes of the operational Army. Compounding the problem is rather fierce and defensive insistence by TRADOC that they were already doing the right things, the right way—even by people who were critical of TRADOC when assigned in the Operational Army.

The central problem is a cumulative effect of *culture*, combined with *organizational design*, combined with changing Army *assumptions* about how to fight and how to train the Army to fight that inhibit TRADOC transformation and mission effectiveness. The Army generally defines culture as a collection of shared values, beliefs, and behaviors. When he was Army chief of staff, General Schoomaker articulated the need to change the behaviors of the Army to reflect new operational realities. He also emphasized adapting an institutional Army that provides soldiers, leaders, organizations, equipment, and training to the Operating Force. The Army must change its cultural frames of reference and move beyond "task/conditions/standards."[12]

First and foremost, TRADOC is a large organization. Although everyone in the Army passes through TRADOC activities periodically, and some spend time immersed in one or two specific activities, very, very few officers or NCOs have deep knowledge of the entire entity. Simply to establish some context to the discussion, consider the following:

> The TRADOC mission is to recruit, train, and educate the Army's Soldiers; develop Army leaders; support training in units; develop doctrine; establish standards; and build the future Army. TRADOC has 11 core functions, 3 key enablers, 3 specified proponents and 16 branch proponents. TRADOC has 37 general officers. There are five deputy commanding generals and five field operating agencies. The TRADOC HQs staff is functionally unique and does not follow the G-staff model. There are six primary coordinating staff departments, with approximately 31 subordinate directorates. The DCSOPS&T alone has 12 sub-directorates. One specified proponent, The Combined Arms Center, itself has 27 functions, 11 subordinate commands, eight subordinate centers, seven schools, and eight special activities.[13]

> If one were to follow the TRADOC model to create a new training topic, concept, or program, this model requires compliance with 22 DA Pamphlets (Pams) and Other Procedural Guidance; seven capstone Field Manuals; 19 HQ TRADOC Regulations and Direct Guidance; and 37 TRADOC Pams and Other Procedural Guidance (averaging 140 pages each).. That does not include any of the topic specific FMs, TMs and other proponent-generated procedural requirements, nor does it include the approximately 32 training related Federal/DoD/Joint Guidance documents. It does include the responsibility to ensure accomplishment of all required communications and coordination, both within the proponent parent organization, with external and higher organizations, and across TRADOC. At a minimum, that would be the three specified proponents and their subordinate activities (particularly CAC), the six TRADOC HQs coordinating staff departments, and many of the 31 directorates all before recommending approval of new policy or guidance, generating a tasking, or beginning to implement a new command initiative.[14]

A discussion of how the Army got to this point is in order, because there are many in the Army who have an almost religious allegiance to the unchallenged doctrine of training to "task/conditions/standards," yet they cannot explain *why* it is the Army way. The impediments to change are easier to counter when the culture is defined as something that can change, and has changed before, without destroying the organization.[15]

Remember, also, that the Army's training responsibility is much more than mechanical, repetitive tasks. Training tactical proficiency also requires training to develop tactical judgment. Current Army behaviors are also not something that must always be defended based on tradition. The warrior ethos is not in the same cultural category as "task/conditions/standards." In other words, culture defined by organizational behavior is not the same as culture defined by organizational values.[16]

The Army's 1973 reorganization of Continental Army Command into two separate commands, TRADOC and Forces Command, roughly coincided with three related events. First was the experience of the Vietnam War and its effect on Army "culture," which was captured in the 1970 *Study on Military Professionalism* and the *Board for Dynamic Training*.[17] Second is the Army's transformation from a conscript to a volunteer (but not yet to a professional) Army. And third are the assumptions that the Army must be able to fight outnumbered and win a major conventional war in central Europe.

The impact of each of these events was strongly influenced by observations from the 1973 Yom Kippur war extrapolated to potential war against Warsaw Pact forces. The importance of these three events is often discounted today by critics who declare that the Army ignored the lessons of Vietnam. There is, however, a very important part of Army culture that did not discount lessons of the past, but does struggle with how to adapt them to new wartime realities, specific warfighting requirements, and service budget program requirements.

General William DePuy, as the first TRADOC commander (1973–1976), had a vision for changing the new volunteer Army into one capable of fighting outnumbered and winning the first battle, while buying time for mobilization of reserve forces necessary for fighting a long and global war. National security strategy, as well as the political realities of the time, cast the Soviet-led Warsaw Pact as both the most dangerous and the most likely threat for the U.S. Army. This Army transformation led by TRADOC in the 1973–1990 period can be characterized by the following doctrine as the unifying engine of change.

General DePuy's personal leadership philosophy became the institutional culture for how to train. DePuy was very much a product of his personal experiences as an operations and commanding officer of a battalion in 90th Infantry Division during Normandy and the drive across France in World War II, and then as commanding general of the 1st Infantry Division during the Vietnam War. In each experience, DePuy was immersed in a 2GW culture. Most importantly, he had experienced draftee soldiers and inexperienced leaders, which fit 2GW culture.[18]

The Army's assumptions about the nature of a conventional war in central Europe were based upon a need for forward-deployed draftees, then, later, volunteers to hold long enough for reservists to be mobilized and deployed. General DePuy's personal experiences with draftees and volunteers reinforced his beliefs about how such forces would have to be prepared to fight that war and win. He believed that there was insufficient time and capacity to develop civilians into professional soldiers. Therefore the institution would have to be top-down and in centralized hierarchies able to "tell them what to do, tell them how to do it, and check that they did it right."[19]

This is an important element because, over time, these assumptions about forward-deployed forces fighting outnumbered, while reserves in the United States went through mobilization that transformed reservists and Guardsmen into combat-ready reinforcements established a culture in which leaders expected to be told what to do, how to do it, and what the standard was by TRADOC. This top-down approach was not intended to negate leader initiative, because General DePuy also believed that leaders must exhibit initiative in order to prevail in combat. But, the scope of the Warsaw Pact threat model and its Industrial Age tradition evolved the U.S. Army to function as a machine, not a thinking organization.[20]

Initiative is hard to define quantitatively, and it does not meet the quantifiable standards or measures that satisfy the peacetime military operations research systems analyst (ORSA) that dominates DoD and DA. ORSAs desire predictability and uniformity and that is what leaders—both military and civilian—look toward to help answer their questions. They then use it as a guide to determine whether they are "doing the right thing" to move up in the system.[21]

DePuy was not wrong, but his ideas were specific to the strategic conditions of his time and to an Army grounded in 2GW. Strategic conditions have changed; the operational Army has evolved, as have the people the Army draws from U.S. society, mentioned in Chapter 3; and threats have certainly changed. However,

Army leaders are just now beginning to re-evaluate DePuy's basic assumptions and the influences of Frederick Taylor's theories on industrial management on how the Army trains.[22]

To evolve the Army culture—personnel management system's hold on recruiting and retention, as well as its training doctrine—Army 2GW cultural behaviors were further influenced by three additional significant events:

- End of the Warsaw Pact as a threat model,
- Army assumptions regarding war in the "information age,"
- Cold War drawdown effect on Army leader behavior.[23]

After the end of the Cold War and Desert Storm, new Army study groups were convened, new doctrinal manuals written, and new operational concepts explored. Army leaders recognized that the single-enemy focus of 2GW would not be relevant to the Army in a 4GW world. As discussed in Chapter 4, Army transformation to a more deployable, adaptive, and agile force began. Information dominance across the tactical and operational levels, enabled by technology, formed the basic assumption about how the Army would fight. The models for designing, testing, and evaluating new concepts remained tied to 2GW types of mathematical and linear threat models used by DePuy to justify force development funding in the 1970s. This model needed definitive assumptions about how the Army would fight that could be rationalized mathematically on the linear battlefield.[24]

Even though the nature of conflicts in which the Army would engage was changing, the Army's assumptions about threat models used to create change did not. After 30 years and victory in Desert Storm, it could not question them without calling into question the basis upon which the size, composition, and required capabilities of the force were justified in terms of budgetary requirements. The evolving face of war into the Fourth Generation through new world situations in which non-state groups were becoming more powerful, as discussed earlier in the book, combined with technological advances on which the Army was spending development funds, led many to conclude that the Army would be able to "do more with less." This is a troubling paradox for a large institution, and its inability to reconcile the desire to be a doctrinally based Army, in a world in which the desired doctrinal measurements cannot be defined quantitatively, persists today. The Army had firmly attached itself to a force development model in which doctrine was not "how we will fight" the nation's wars, but "how we will justify acquiring and managing resources" on a macro-level. Doctrine no longer was the engine of change, because the extensive bureaucratic systems that were built in the post–World War II 2GW world now held doctrine captive to process.

Not only did assumptions about how the Army must be developed not change, but many 1970s' assumptions about how the Army should train remained unchallenged. No one would now argue that the accepted methods used to train soldiers for World War I did not apply to the new battlefield realities of World

War II a mere 23 years later. But oddly enough, there are many in the Army today who passionately argue that methods developed to train for winning a first battle in central Europe in the 1970s and 1980s are still unquestionably valid for the current war in 2008.

In fact, the success of Desert Storm exacerbated the problem by apparently validating DePuy's training philosophy of "task/conditions/standards," which no longer was recognized as his philosophy, but instead firmly embedded as Army culture. Leaders raised under that philosophy chose not to question it—even in the light of Somalia, Bosnia, Kosovo, and 4GW around the globe. Tactical problems were viewed as either the failure of subordinates to understand doctrine, failure to develop detailed standard operating procedures for applying the doctrine, or political failures resulting in the improper (non-doctrinal) use of the Army, certainly not indicative of a need to critically examine Army assumptions and doctrine. Mission Essential Task List (how commanders determine training priorities based on their analysis of the missions they may have to conduct) development has evolved from a tool for commanders to focus efforts to their mission, to a very broad listing of approved task words related to larger and centralized higher headquarters' use of certification as the primary method for assessing readiness.[25]

While the direct connection between how to train doctrine and how to fight doctrine was unhinged by the emergence of 4GW, Army leaders' attempts to retain 2GW training doctrine remain strong. To challenge the rationale of the methods used to train was to challenge the doctrine itself, which was a professional behavior increasingly less desired by an Army that celebrated the leaders and successes of Desert Storm.

A concurrent event, unrelated to war-fighting, helped entrench organizational conservatism upon the Army—the dramatic drawdown of the 1990s, in which the Army shrank by half. The effect of the drawdown on people who remained/survived was to instill a strong professional conservatism and groupthink. In the 1980s, often public Army internal debates about how to fight and how to train that had accompanied evolving Army doctrine were supported indirectly by robust resources. Even with DePuy's powerful philosophical influence, there was still the flexibility to experiment, improvise, and exercise command initiative. This was especially true in United States Army Europe and in first-deploying forces, where units were closest to enemy realities, and also where operational commanders had much more practical influence on their subordinates' thinking than did TRADOC policies. Army internal debates certainly included TRADOC but, interestingly, the adaptive concepts for light infantry divisions and the high-tech test division in the 1980s were directed by the Chief of Staff of the Army (CSA) prerogative, not TRADOC initiative.[26]

The abundant resource environment in the 1980s allowed robust debate about, and innovation within, Army doctrine. The soldiers and officers who participated in small-scale combat, as well as those who conducted mission-focused training in the Combat Training Centers (CTCs), Germany, and Korea developed into professional, not just volunteer, soldiers.

The drawdown and resource crises of the 1990s slowed and almost stopped innovation. Resources, and thus training, became increasingly centralized. Junior leaders could not be allowed to squander limited resources learning their craft; they needed to be taught "what right looks like" by their seniors, because there was too much risk in allowing junior officers and NCOs to develop professionally the same way their seniors did. Innovative training methods such as those employed by Special Forces were considered inappropriate for conventional forces. Junior leaders emulated the behaviors of their seniors, centralizing and directing the "task/condition/standards" activities-driven subordinate activities, and held doctrinal correctness as an essential measure of leader competence.[27]

Training as a leader competency was also replaced by training resource management as a leader requirement quite unrelated to the actual execution of training. Careful stewardship of resources, and the satisfactory completion of resourced events, culturally took precedence over the actual effectiveness of training. Training itself changed from experiential training (proficiency gained through experiencing realistic training, incidentally a key part of DePuy's training philosophy) to event-driven training, following strategies determined by TRADOC. These strategies determined the approved methods and allocated resources and external "trainers" for unit commanders. The CTCs changed from an environment in which leaders trained their units to fight, to a place where outsiders told leaders to follow approved doctrinal methods. The fundamental training methods remained unchallenged by critical analysis and became the hallmark of leader risk mitigation. Leaders that survived the drawdown ended up following doctrinal methods precisely and evaluating others by how well they followed the same methods.

In 2000, the gap between intentions and reality became public as Army officers and NCOs began voting with their feet. The volunteer professional Army was losing staff sergeants, captains, lieutenant colonels, and colonels faster than even a smaller Army could handle and still fulfill its requirements. Command climate surveys showed wide and deep dissatisfaction with senior leaders, with Army schools, with training methods, and with overly restrictive command climates. Highlighting the disconnect, it is interesting to note that while General Eric Shinseki, Army chief of staff from 1999 to 2003, commissioned a series of Army Training and Leader Development Panels to try to understand why the gulf existed, the TRADOC commander continued to insist (in testimony to the House Armed Services Committee (HASC) that training and leader development were the strengths of TRADOC and that there were no issues requiring fundamental change.[28]

By 2003, with the Army CSA, General Schoomaker, engaged in changing the Army, not only to help win an ongoing war but to prepare for future security needs, still found himself hamstrung by a generation of subordinate leaders in the institutional Army who survived and thrived by not changing any systems without first being told the approved answer. Army leaders—officers and NCOs—became victims of goal displacement. Faced with uncertainty and ambiguity, they transformed what their cultural experience told them they could do (and could not do), into what they believed they should do. It only became worse when the broader organizational and senior leader culture did the same, misusing methods such as quarterly training briefings to measure events as if they were measures of effective training.[29]

Until recently, the Army was using training methods for wartime readiness as if it were still the 2GW Volunteer Army of General DePuy's day, expecting to be told what to do, how to do it, and what the standard is. Everyone was assumed to know nothing until trained and certified by an outsider. Indeed, in the operational Army there is reliance on TRADOC doctrinal products (and the hierarchy of multitudes of quality assurance inspectors) as a substitute for commanders' vision, concepts of operation, and innovative training strategies.[30]

On the unexpected eve of 9/11 and the following campaigns in Afghanistan and Iraq, the Army found itself with new operational requirements and many new technologies, but a professional officer and NCO corps that did not really know how to train it to fight and wage 4GW. It could execute published tasks, under defined conditions, to simulated standards, but the culture imposed upon the Army caused it to struggle with how to fight and how to train to fight when conditions or requirements did not conform to officially approved assumptions. The Army had, through its own cultural behaviors in the 1990s, taught itself what to think, not how to think.[31]

Colonel Casey Haskins, commander of the Army's 198th Infantry Brigade, responsible for the training of new Army infantry soldiers at Fort Benning, GA, describes the result in an observation he made in his own unit in August 2007:

> The training program lacked focus and often seemed to drill sergeants and company leaders like a non-stop series of events without connecting themes or clear priorities. A company commander trying to manage all the training could easily feel like the ball in a pinball machine, bouncing from one event to the next. The printed documentation governing 14 weeks of training was a stack of papers over three feet high, virtually unreadable.
>
> Much of the training had an assembly line, check-the-block feel to it. Soldiers firing on a machinegun range would move forward, pull the trigger, and then be moved off the line. They did not leave the training feeling confident they knew what to do.[32]

Colonel Haskins contributes this observation of training to three factors that evolved from DePuy influences which led to greater centralization and a top-down Army culture:

- Outsiders, usually trainers from 2nd Battalion, 29th Infantry (2-29), delivered much of the training. This left the company chain of command and especially the drill sergeants with little sense of ownership and little vested interest in seeing that the training was effective.
- The second factor was the low expectations embedded in the TSPs [training support packages], the documents governing what was to be trained and how. Not only did

the TSPs often assume that Soldiers were capable only of achieving minimal standards, but they sometimes dodged the issue altogether, requiring only "familiarization" with a topic—a vague term usually defined down on the ground to require mere presence and rarely requiring anything rigorous.

- The third systemic factor discouraging excellence was that training was governed by inputs rather than outcomes or results. Number of hours spent, number of rounds fired, number of classes attended all counted; true proficiency at the task being trained seldom did. This system made it easy to calculate resources and provide support but, unsurprisingly, did not maximize the training effect. Attendance at land navigation training was a prerequisite for graduation; the ability to read a map and navigate from one point to another was not.[33]

As a result, recent leader and soldier training do not encourage thinking and decision making. In fact, it often discourages it. Although the best instructors—and especially those recently returned from combat—take great efforts to explain to their soldiers why things were done a certain way, the program itself stressed only the mechanical application of tasks. Worse, the atmosphere established during some courses emphasized "total control." In some units, particularly basic training units extended beyond the point of usefulness, that atmosphere sometimes remained nearly until graduation. Drill sergeants yelled, while instructors at leader courses assumed the "know and be all" stance that prevented anyone from questioning their authority. Cadets, candidates, and junior officers, as well as soldiers, asked few questions, and infractions were answered by mass punishment, while education techniques are rote and boring.

Young leaders and soldiers are not forced to work things out for themselves or to learn to be individually responsible. Not understanding why tasks are performed a certain way, they often fail to adapt properly to changed circumstances. Fortunately, thousands of leaders at the officer, NCO, and retired levels in the Army have recognized the downfalls of today's training and education doctrine and are moving from the bottom up to fix it, better preparing tomorrow's Army for the changing face of war.

Today's Training Revolution

> "Training for warfighting is our number one priority in peace and in war. Warfighting readiness is derived from tactical and technical competence and confidence. Competence relates to the ability to fight our doctrine through tactical and technical execution. Confidence is the individual and collective belief that we can do all things better than the adversary and the unit possesses the trust and will to accomplish the mission."
>
> Army Field Manual (FM) 7-0[34]

More than ever is being asked of our junior leaders and soldiers—today's *strategic corporals* and *lieutenants*. Existing knowledge centers (academic centers) provide an opportunity to take advantage of the learning environment of college campuses, while "massing" more expensive tangibles to train tasks at Army bases. Leaders—officers and NCOs—are seeing what needs to evolve and are now drawing on a wealth of theoretical as well as practical experience and are beginning to field and lead a new training revolution.

Behavior changes are not lasting if we fail to strike at their antecedents. Until relatively recently, these causes were not well understood, so there was little the Army could do to influence the way they developed cadets in meaningful ways. This began to change based on research done since about 1970. Today, the Army must start thinking in Quantum, rather than Newtonian, terms with regard to leader development. That is to say, the Army has to account for the fact that the actions we take at the earliest points in a career and thereafter, in a sequential and progressive fashion, manifest themselves much later.

In order to counter an array of national threats and opponents, using practices that range full spectrum, a synthesis of Army courses into "Learning Organizations" is needed. To meet this educational end, current educational and training ways and means must be assessed, evaluated, and changed. Weak spots and points of failure in leader and soldier education and training must be identified—all in the interest of retooling the system in ways that facilitate the development of officers who are intuitive and adaptive.

The Army is currently embracing two new and evolving (not revolutionary) approaches to leader development and soldier training, while there is also an excellent example of a commander's initiative for change and implementation of it. They are

- The Combat Applications Training Course (CATC),[35]
- The Adaptive Leaders Methodology (ALM),[36]
- 198th Infantry Brigade efforts at Fort Benning, GA.

The Army's Asymmetric Warfare Group (AWG)[37] that teaches the CATC philosophy deals with a method to instruct and develop mastery of any given subject. Its premise is that soldiers can apply principles to understand the how and why of training. At the center of CATC is the use of problem solving in order to teach a task. In this venue, CATC uses rifle marksmanship as the training vehicle. CATC has been taught to several battalions throughout the Army as well as at the drill sergeant's school at Fort Jackson, SC, and infantry training brigades at Fort Benning, GA.[38]

Several unit leaders attest to the effectiveness of the CATC approach to soldier development. Captain Micah Niebauer of 82nd Airborne Division; First Sergeant Rick Skidis of B company, 1st battalion, 502 Airborne Infantry, 101st Airborne Division; Command Sergeant Major Brian Carlson, Command Sergeant Major of Fort Jackson, SC (Army Training Center); Colonel Michael Coss, commander of the 192nd Infantry Brigade, Fort Benning, GA; and Colonel Casey Haskins, commander of the 198th Infantry Brigade, Fort Benning, GA, have asked for and now support the CATC in their units. It is used for both operational and basic and advanced individual training, and, most importantly, they testify that the course is better at making their soldiers professionals who are prepared to fight and wage 4GW.[39]

During CATC, teaching a military skill results in mastery of that skill for the vast majority of students, when presented as a relevant problem-solving exercise. In this manner, the teacher guides the student to discovery of a desired solution or outcome through established principles or demonstrated facts, and with student-generated evaluation throughout the learning cycle. Mastery of skill results in

- Confidence, accountability, and initiative,
- Introduction to the skill of discursive reasoning and the norm of problem solving,
- A tangible skill.

AWG instructors believe this is applicable to all manner of military training and its efficacy has been readily demonstrated in history and in current military initiatives. The expression "mastery" defines reasonable ownership of a skill in terms of knowledge, expertise, and application. The principles of CATC are these:

- *Training to Grow Problem Solving* teaches soldiers to "teach themselves" the skills necessary to the success of their mission.
- *Training to Increase Intangibles* develops the intangible attributes of confidence, accountability, and initiative.
- *Training to Increase Understanding and Awareness* teaches through contextual understanding of the task and its mission application.
- *Training to Increase Deliberate Thought* conditions soldiers to always exercise a deliberate thought process while under stress.
- *Training to Improve Combat Performance* conditions soldiers to overcome the psychological and physiological effects of combat.[40]

The CATC program uses rifle marksmanship as a foundation to improve the professionalism—confidence and character—of the soldier. An example CATC that illustrates the difference between Industrial Age training (negative) and training professions (positive) is how to clear a soldier's weapon, called "SPORTS" (slap, pull, observe, release, tap, and shoot).[41]

The Industrial Age approach (negative)—only presents task, conditions, and standards, so the problem for the soldier is correctly performing a mechanical function in 5 seconds, without the problem-solving opportunity of considering the situation of a jammed weapon in combat.

This is different from the professional (positive) approach, which is to present a mental situation of expected combat with a weapon that is jammed, so the soldier begins to think about what can be done, then talk through the logical assessment and clearing of the weapon with SPORTS, so the soldier performs a thoughtful task in the expected 5 seconds.

The following is a short narrative for the benefit of instructors or teachers. It is a natural process for individuals to solve problems that have reached a point of concern. If an individual is in a position of influence or authority, and identifies a problem of concern within an organization, a solution is generally presented in the manner demonstrated by the following three examples.

Soldiers need to be able to clear rifle stoppages. This can be done by performing SPORTS. Soldiers must perform SPORTS when they encounter a stoppage. Soldiers must be able to perform SPORTS correctly in 5 seconds. They are supervised in their training by following the process, while instructors (drill sergeants in most cases) do the task for the soldier. In effect, this takes the responsibility out of the hands of the soldier when they perform the task themselves. In turn, this makes the soldier dependent on the instructor and, later, the chain of command for trivial tasks.

Mechanics need to be protected from electrical shock while working on an engine. This can be accomplished by disconnecting the battery leads. Mechanics must disconnect the battery leads before working on the engine. Shop managers will ensure battery leads are disconnected by the mechanics prior to working on the engine.

Assembly line workers need to be able to clear debris jams on their section of line. This can be accomplished by halting the line, releasing the track pressure, removing the debris, resetting the track, and restarting the line. Assembly line workers must perform this procedure when they encounter a debris jam. Assembly line workers must be able to perform this procedure correctly in 30 seconds. Factory managers will ensure the procedure is followed by the assembly line workers prior to and as they are working on the assembly line.

These examples contain four key elements.

- The stated problem,
- The stated solution,
- The action that must be taken for success,
- The means to ensure the action is or will be performed correctly.

It can be argued that there is nothing systematically wrong with this manner of presenting solutions to identified problems of concern: It is to the point, it appears efficient, and it is quantifiable. The Army, and society, have a tendency to solve organizational problems using the mechanical approach or to find a process, such that the solution is thrust upon the group with a means of enforcement. The focus of this type of problem solving is the problem itself, and it often relies on principles of manipulation associated with models like Maslow's hierarchy of need.

CATC developers have discovered another approach to problem solving, but it is counter to current cultural ideas and is not readily inherent in society. It is extremely powerful in building a sense of accountability, responsibility, and confidence in the soldier in addition to solving the problem. It is especially suited to volunteers. The focus of this type of problem solving is at least equally divided between growing the intangible attributes of juniors and the problem itself, and relies on the principle that individualized, purposeful action is superior to mandated action.

In this approach, CATC instructors have discovered when an organizational problem is encountered, a solution is still defined (to a degree), but it is not directly presented. The leadership works with the soldier to help him see and analyze the problem from an operational perspective. The leadership demonstrates to the soldier that the resources required to solve the problem are within the soldier's means. In this manner, the soldier plays an active (creative) role in discovery of the solution. Two things are accomplished with this approach: The solution is found through the positive experience of discovery with the soldier taking ownership of the solution, and the soldier experiences the process of problem solving from an operational perspective. Revisiting the simple example of a rifle stoppage, there is the following example:

"An obstacle in your mission to engage the enemy is a stoppage in your rifle." How can we overcome this obstacle? "First, let us determine what causes a stoppage and then how to correct it." The leadership talks out the matter, using its defined or partially defined solution as a guide—that is, the logical and thoughtful process of SPORTS versus a mechanical function that must be performed correctly in 5 seconds. The intent is for the soldier to discover the cause and solution by his own deliberation. This builds in the soldier a sense of accountability for his actions, responsibility for taking action, and confidence in deciding on the correct action. The goal in this approach is for the soldier to understand the problem and the solution. The standard of correctly performing the task in 5 seconds will be a natural by-product and, to a degree, is secondary to the growth of the soldier.

Asking questions in a positive manner involves the soldier in the process of discovery. Using the word "we" means the leader is joining with the soldier in finding mission success for the soldier – "this is your mission (or task), but I am committed to helping you find success."[42]

Contrast the two methods of dealing with organizational problems—Industrial Age and Professional—and discover that the first method is not concerned with growing the problem solving ability of the soldier. It also focuses the efforts of leadership on solving an increasing number of problems for the soldier. The by-product of this method is a growing sense of frustration in the leadership, over the growing sense of a lack of initiative and responsibility in soldiers. The second method is concerned with growing the problem solving ability of the individual and sees problems as opportunities. It focuses efforts of leadership on providing the environment and resources for the individual to succeed. The first method defines what the soldier *must* do for success while the second method empowers the soldier to discover what he can do.

"It is better to be called a mentor than an instructor."[43] An instructor states the solution (learn the skill being taught) and then states the problem (standards the soldier *must* meet). Notice how the problem for the soldier has shifted from the need to clear a rifle stoppage in combat to the need to perform a task correctly in 5 seconds. This places a requirement on the student not related to the task of war-fighting and contrary to the idea of volunteerism found in American soldiers. A mentor helps the individual understand the problem as it relates to the task of war-fighting and then guides the individual to discover (a creative process) the solution (mastery of the skill). A mentor does not have to know the solution; he only has to ask the question and be willing to help find the solution.

Field Manual 3-22 Rifle Marksmanship contains the word *must* more than 400 times, with more than 100 references to *Soldier(s) must*... By doing this, the Army has solved the problem for the individual and retains ownership of the solution. In doing so, it also removes operational accountability, responsibility, and confidence from the individual. Leaders and advisors may find it rewarding to solve problems—we all like to do it—but if we are not teaching soldiers how to solve their own problems and growing an environment where they are resourced for success, then we are the problem.[44]

Can we see the difference between a conscript, a factory worker for hire, and an American soldier—enlisted, NCO, or officer—as long as the Army remains stuck in 2GW? Fortunately, many of the Army sees the answer as yes with regard to how the Army trains and develops its leaders and soldiers. Parallel to the CATC program of instruction is another innovation that evolved from an effort to develop cadets to be better decision makers and leaders of character at Georgetown University ROTC between 1999 and 2005. Outlined as the Adaptive Course Model (ACM) in the book *Raising the Bar: Creating and Nurturing Adaptive Leaders to Deal with the Changing Face of War*, the leader development approach's name was changed twice after I came to work for Army TRADOC at Army Capabilities Integration Center Forward (formally Future's Center). Not surprising, while CATC and ALM use different vehicles—ALM uses tactics—to develop professionalism and decision-making skills, their methodology is similar.

Warfare has evolved to the point that the central idea is that small unit leaders in direct contact with the enemy can see and react to situational changes much faster than could the more senior leaders in the rear. This occurs despite the advent of information technology. This technology laid over an Industrial Age hierarchal force structure can make it tempting for leaders to micromanage. Thus, the decision cycle slows down. Small unit leaders who were once only concerned with choosing which battle drill now make decisions that have strategic implications.

The Army acknowledges the need for change. The Army has begun an evolution in the way we develop—train, educate, access, promote and select—leaders, specifically how do we evolve adaptability. In the past, the "competency theory" of learning dominated course curriculums, and there remain signs of it continuing today in leader development. A good example of competency theory is "leave no child behind." It prescribes what to think and not how to think—"teaching the test." Order and control are central to Programs of Instruction (POIs) that use the "competency theory" as its foundation.[45]

Leader development for asymmetric war must be based on quality, not quantity, at every grade level. The rule should be, "Better no officer [leader] than a bad officer [leader]."[46] Schools must constantly put students in difficult, unexpected situations, and then require them to decide and act under time pressure. Schooling must take students out of their "comfort zones." Stress—mental and moral as well as physical—must be constant. War games, tactical decision games, map exercises, and free-play field exercises must constitute bulk of the curriculum. Drill and ceremonies and adhering to "task, condition, and standards" (task proficiency) in name of process are not important. Higher command levels overseeing officers' and NCOs' schools must look for courses adhering to a few principles, while allowing instructors to evolve their lesson plans using innovative teaching techniques and tools to an ever changing environment. Those leaders who successfully pass through the schools must continue to be developed by their commanders; learning cannot stop at the schoolhouse door.

Current research—the work of Dr. Robert Bjork at UCLA— tells us the most frequent type of decision making for leaders in a time critical environment is recognitional, which requires a large amount of experience. Research also tells us that competence in decision making is solidified by making a large number of decisions in a stressed environment. Leaders must understand that deciding when and how to close with an enemy may be the least important decision they make on an asymmetric battlefield. Instead, actions that build and nurture positive relationships with a community, local leaders, and children may be the defining factors for success, as well as the primary tools that contain an insurgency, build a nation, or stop genocide. True tactical prowess often entails co-opting the local population's will while shattering the cohesion of asymmetric adversaries.[47]

Educating a future leader in "how to think" (cognitive skills) takes longer and is intellectually far more expensive than industrially based task training, while task training requires resources like weapons, ranges, equipment, and special facilities that require training be done at established locations, requiring centralization. The good news is that recent studies by Dr. Robert Bjork, dean of the School of Psychology at the University of California Los Angeles (UCLA), have discovered that theories about learning have been wrong. The focus on process, training for the test, and task proficiency benefits short-term training but does not promote long-term learning. The translation for Army education is that task training can be done anytime in an officer's development, but the longer you wait to develop cognitive skills, the harder it becomes.[48]

Bjork's work, as it relates to evolving the current task-centric and process-centric approach to Army education, can be summed up in the following two statements:

- "Conditions of instruction that make performance improve rapidly often fail to support long-term retention and transfer, . . . *whereas*
- Conditions of instruction that appear to create difficulties for the learner, slowing the rate of *apparent* learning, often optimize long-term retention and transfer."

Army courses using ALM program of instruction (POI) expose students to classical education in conjunction with existing leadership programs on campuses where they are taught to find the answers, whereas current curriculum as described earlier in what CATC discovered gives soldiers the answers. Instead, if the student is exposed to an environment in which they want to find the answers for themselves, then the lessons are emotionally marked in time, which builds intuition—a necessary trait of "adaptive leaders." This approach in leader development (ALM) immerses the student in education and training with innovative teachers. This is the same methodology mentioned in Chapter 1, "Tomorrow's Battlefields?," that prepared Lieutenant Pat Fagan and many of his peers for 4GW.

The ALM is a new leader development model incorporating recent advances in the field of experiential learning. It is an answer to the call for a new leader education model to reshape a fundamental Army learning process for a dynamic operating environment. The ALM provides an answer to the 2006 TRADOC Area of Interest 2's call for "Change [to] the Professional Military Education (PME) model to adapt to the contemporary operational environment (COE) and the Army Forces Generation (ARFORGEN) model, and leverage Army Distributed Learning (ADL), which supports Army TRADOC Campaign plan."[49] ALM was mentioned—as the Adaptive Leaders Course (ALC)—to the commanding general of TRADOC, General William Wallace, under the category of "What Works" at the TRADOC Science of Learning Workshop as part of the Leader Development working group outbrief to him in August 2006.[50]

The Basic Officer Leader Course (BOLC) II courses at Fort Sill, OK, and Fort Benning, GA, have been using ALM since July 2006. During the period of February 2008 through August 2008, the demand for information on ALM intensified as well as requests for the workshop "Deciding under Pressure and Fast" that teaches ALM. Since January 2008 trips to brief ALM and conduct the workshop have been to San Diego, CA (Joint Conference on Military Ethics); Fort Huachuca, AZ; Fort Benning, GA; Fort Monroe, VA; Fort Knox, KY; and the United States Military Academy (USMA) at West Point, NY.

ALM has become institutionalized as Lieutenant General Benjamin C. Freakley, Commander U.S. Army Accessions Command, signed a policy letter, dated 24 April 2008 titled "Basic Officer Leader Course (BOLC) Policy and Guidance," mandating ALM certification for BOLC instructors. The Asymmetric Warfare Group (AWG) has also used ALM in its incentives, and it hosted an Adaptability Conference 3–4 June 2008 with Day 1 focusing on ALM's workshop while Day 2 focused on "Outcome Based Training."[51]

Many other institutions within the Army, including leader-centric courses and operational units, have also used ALM as an approach to improve leader development and soldier professionalism. This supports the CSA and SA vision supporting the need that the Army adapts its culture to encourage, develop, and teach adaptive leadership. The issue before TRADOC is instituting a process that moves beyond this vision to a tangible method to instruct our leaders "how to think" versus "what to think."

Already, cadre leaders at BOLC II are using ALM principles. These principles include

- Providing "contextual interference" during learning (e.g., *interleaving* rather than *blocking* practice), most tasks become learned through doing, and are subordinate to leader development scenarios, taught as needed and as part of varying the conditions of learning.
  - "Experience the thing before they try to give it a name."
  - Conduct scenarios three levels higher to understand their (units') role in the bigger picture through use of tactical decision games (TDGs).
- Execute free play force on force at platoon level (w/PLT TAC acting as company commissioned officer), with missions used as vehicles to develop leadership adaptability.
- Distributing or spacing study or practice sessions (provide the opportunity and access to find answers).
- Reducing feedback to the learner, forcing the student to find the answer (less telling them the answer).
- Using observations and evaluations through conduct of scenarios (rather than presentations) as learning events.

Their feedback and that of students and others reflects the positive impact this cultural change has on our Army's future leaders. Of significant note is that this "change" has required no additional resources or a lengthening in the total period of instruction. While the ALM takes advantage of our current combat veterans' insights and experiences, it requires their continued initiative and desire to train and help grow future leaders. The ALM does this because it continues to build on the Army core principles and values. The warrior ethos underpins everything in the ALM, while the ALM adapts our Army's leaders to the current and future operating environment.

Adaptive Leader Methodology is a cultural change rather than a specific set list of exercises. ALM develops adaptability through the Rapid Decision-Making (RDM) process using the experiential learning model through scenario based learning. ALM is a system that promotes self-actualized learning via weakly structured situational problems. Additionally, ALM parallels the latest findings of the academic world in leader and cognitive development. The ALM program of instruction (POI) employs techniques that are "desirable difficulties" as pointed out by Dr. Robert Bjork in his keynote presentation at the TRADOC hosted "Science of Learning Workshop" August 1, 2006. ALM espouses institutionalized inductive reasoning in order to prepare leaders for the complex wars of the future.[52]

At a course using ALM, students are quickly thrown into problem solving exercises that would be viewed in the past as "too complicated for them without first learning the basics (from a classroom lecture)." They then review the results of their actions in an after action review (AAR) in which the instructors facilitate the students in finding their answers. The instructors avoid telling the students how to do it, there are no book solutions, but they guide the students toward workable solutions they already discovered in experimenting during the course of the scenario. Preferably, the instructors use force on force, free-play exercises whenever possible, but in lieu of these capstone exercises, the instructors use TDGs as a tool to facilitate learning, or symposium-based case studies, before ever introducing theory or doctrine (it is desired the students discover these on their own).[53]

Students are allowed to run as much of the course as possible. If it is rifle qualification that is required as part of the course curriculum to meet Army regulations, instead of using the Industrial Age method, students not only learn how to shoot, but since ALM is now the POI of a leader-centric course, they also learn by running the range as well. These approaches do not alleviate the cadre of the responsibilities of teaching and ensuring the safety of the students; and as a matter of fact, the approach called for in the POI is more difficult because the instructors must stand back and let the students learn through doing, but also know when to step in to keep students on course without wasting too much time as some student leaders will flounder in trying to lead and solve the problem.

The ALM holds to the idea that every moment and event offers an opportunity to develop adaptability. Every action taken by a student in the classroom or in field training is important to the process of inculcating a preference for new solutions. If students err while acting in good faith, they do not suffer anything more than corrective coaching. Constructive critiques of solutions are the norm, but more important are the results of actions, and the reasons for those actions. The role of coaching and 360-degree assessment is to develop students so their future actions will make a positive contribution to their unit's success, no matter what the mission. This idea is based on the premise that one learns more from a well-meaning mistake reviewed critically and constructively than from applying an established and memorized process.

ALM teachers will be very concerned with why the students do what they do—an action-learning approach. The emphasis of the course will be on ensuring that the students gain and maintain a willingness to act. During numerous AARs and mentoring sessions—occurring during and after numerous scenarios with different conditions—the teacher will analyze why the students acted as they did and the effect the action had on the overall operation.

The ALM curriculum and leader evaluation system will use two criteria to judge whether students did well: the timelessness of their decisions and their justification for actions taken. The first criterion will impress on students the need to act in a timely manner, while the second requires students to reflect on their actions and gain insights into their own thought processes. Since students must justify their decision in their own mind before implementing it, imprudent decisions and reckless actions will be less likely. During the course, student decisions in terms of a "school solution" will be relatively unimportant. The emphasis will be on the effect of the students' actions, not on the method they may have chosen. This encourages a learning environment in which there will be few formulas or processes to achieve optimum solutions. This environment will solicit creative solutions.

The learning evaluation system in the ALM is based on the philosophy that feedback should be given in a way that encourages a willingness to act and then reflect on actions in a manner that maximizes learning. Unconstructive critiques destroy the student leader's willingness to act and can lead to withholding of adverse information or false reporting. The course will avoid formulaic solutions and provide room for innovative solutions in its POI. This begins at the entry level to achieve transformation over a generation of leaders, teaching new dogs new tricks.

While the efforts of Colonel Haskins and the 198th Infantry Brigade appear to involve CATC and ALM, they did not.[54] It is another example of a bottom-up effort of innovative leaders of character who see a problem, then take on the established way of doing things in order to fix the problem regardless of the risk to their own careers. They know they are right. In the end, Colonel Haskins' efforts improved the way his brigade prepared soldiers for combat. Colonel Haskins lived "teaching old dogs new tricks every day" by pushing up his chain of command and changing the culture his subordinates had accepted as gospel.

Colonel Haskins began his reforms by realigning responsibility with authority and holding leaders accountable for results. He told his chain of command that they must ultimately be responsible and accountable for all aspects of training. But he assisted them by aligning outcomes, not process. In turn, this aligned the need for standard, high-quality training with room to innovate and improve.

Colonel Haskins realized that what was required was to "unfreeze culture" and, just as important, to redevelop his leaders. He began with his company commanders and first sergeants by teaching them how to plan, prepare, and assess training. Colonel Haskins also pushed for all of the brigades' drill sergeants to become experts who could perform all tasks to standard. They could explain why tasks were performed that way and, finally, they could teach the tasks.

As a good leader, Colonel Haskins also realized his changes needed constant assessment to evaluate their impact and to make adjustments so they could continue the evolution and changes he was seeking. He developed true assessments of outcomes, including all-important intangibles. These results must drive changes and be used for holding leaders accountable. And finally, supporting systems must be made to support instead of impede.

An example was the Army's infamous Range Control, responsible for the safe conduct of all ranges on an Army post. On some posts, Range Control was an authority in itself, with incredible power to shut down training at the slightest infraction. It was feared by many leaders because no matter how good their training was, an inspector from Range Control was sure to find something wrong and would shut down the range, and of course report the infraction to that leader's chain of command.[55]

Colonel Haskins put the training back in the hands of his leaders and negotiated and discussed with his higher headquarters, as well as those who oversaw Range Control, the specific responsibilities of Range Control inspectors. He moved to diminish their power of inspecting training and safety and to shift that authority and responsibility to leaders within his brigade. He describes how hard it was to change this in the following story regarding changing a simple policy of "rodding weapons":

> The first attempt to change an outdated rule was ceasing to use rods to clear weapons after firing on a range. This decades-old procedure prevented mishaps but also prevented Soldiers from learning to be responsible for handling their own weapons safely. It failed to prepare them for combat, and it prevented us from fully meeting the intent of the weapons immersion program, explicitly designed to reduce fratricide. This was a simple change, but it took many months and was fiercely argued, with lots of emotion on all sides. Range Control, G3, the Fort Benning Safety Office, and many others opposed it and fought to prevent it. Within the brigade, many senior NCOs expressed grave concerns, fearing that this would result in fratricide and career-ending consequences. It took months to work through all these issues, and it was only with the Deputy Commanding General's strong support that we succeeded in getting a better system in place, one piece at a time. Now, a year later, it is easy to look back and see that we are actually safer during training—vastly fewer negligent discharges and far fewer instances of dangerous weapon handling—and that we are doing a much better job of preparing Soldiers for combat—they leave here responsible and knowing how to handle loaded weapons safely.[56]

It was leader adaptability that he moved toward and he did this by making himself the role model in how his leaders should evolve and accept a better way to train new soldiers.

Colonel Haskins increased the emphasis on leader training. It was here, as a commander, he had to devote most of his time to change behaviors:

> Changing this required a significant investment. I developed a two-day leader-training seminar and required all captains to attend. The majority of it I taught myself, with assistance from senior leaders in the brigade. A mixture of classroom lectures (few), discussions (more), and outdoor hands-on training (the majority), the program aimed to accomplish three things: to make sure that every leader understood and could clearly articulate the outcomes-based approach to training; to provide company commanders with some very useful techniques and tools; and to cause them to meet each other, share ideas, and have fun.[57]

He focused on developing disciplined, flexible problem solvers rather than on the rote application of prefabricated solutions, as he describes subsequently:

> More fundamentally, I wanted to gain enthusiastic converts, to make sure every company commander thoroughly understood my intent, and I wanted to prove to them not only that training could be enjoyable, but that fun training was frequently more effective. We delved into practical training management,

> showing them ways to use their cycle plan and training meetings to accomplish what they wanted by prioritizing and focusing on the most high-payoff events, and by decentralizing a lot of training to platoons, using assessments to hold them accountable for results.[58]

Colonel Haskins did this by ensuring each leader was proficient at executing every task, and the leader understood why the task was performed that way. This in turn developed teaching skills, and it also was enhanced by re-teaching company commanders and first sergeants how to plan, prepare, execute, and assess training. He also gave his battalion and company chains of command more authority to conduct training the way they wanted, but held them much more accountable for results. He worked with the post to work more flexibility into how the range or training is conducted (task, purpose, and method), while also providing companies more training aids for longer periods of time to enable more decentralized training. He also used more effective and continuous assessments throughout cycle to allow adjustments, if needed. But, most importantly, he encouraged and relied more upon leaders' judgment and less upon checklists.

Conclusion: Evolution Must Continue

> "Most Army schools open with the standard bromide: 'We are not going to teach you what to think ... we are going to teach you how to think.' They rarely do. Critical thinking is both art and science. There are techniques to critical thinking, such as careful application of logic, or alternative application of deduction and induction. These techniques can be taught and learned."[59]
>
> Brig. Gen. David Fastabend and Robert Simpson, "Adapt or Die"

Critically important to the institutionalization of adaptability, which will assist with recruiting and retaining good soldiers in the Army, will be superior (innovative) military education and training. Not only will the Army need to produce leaders who possess adaptability, but the institutions tasked to develop leaders will need to become adaptive as well—to evolve as the future operating environment evolves. The CATC and ALM, as well as Colonel Haskins' move to change the training culture, are examples that provide principles that allow implementation of central ideas.

The Army's cultivation of adaptability requires a vast effort—from the top down as well as bottom up. It is so central to the future of the Army that it applies to squad leaders as well as to the joint-force commander. The leaders of the future Army should have to make a truly gross error to create a negative blotch on their careers. Evaluations and performance reviews cannot continue to haunt adaptive leaders throughout their careers if they have only made an honest mistake.[60]

Moving the Army toward a learning organization structure as Colonel Haskins did with the 198th Infantry Brigade, in which its institutions as well as its leaders are adaptive, will bring the collective creativity of the Army to bear in solving problems at the tactical, operational, and strategic levels of war, as well as in problems that involve recruiting and training. The culture will become one that rewards leaders and soldiers who act, and penalizes those who do not. Today's culture needs to evolve so that the greater burden rests on all superior officers, who have to nurture—teach, trust, support, and correct—the student who now enters the force with the ability to adapt.

The Army's future leaders will also have the responsibility to self-police their own ranks, particularly early on if they become teachers within a CATC, an ALM or an operational unit. This makes evaluating, "racking, and stacking" of graduates easier. It will also help determine early on who will have the character and traits to become an adaptive leader. The criteria should include observations of the student leaders in several scenarios. Before selecting or promoting subordinates, a commander and teacher should always be asked, "Would I want this person to serve in my unit?" Throughout a course, an instructor or teacher will instill in students the importance of accurate reporting and taking action when the situation demands it. The Army's culture of the future will not tolerate inaction. Indecisiveness or the inability to make a decision will become the culture's cardinal sin, not playing it safe.

Adaptability will become a product of the future Army; it will depend on what appears to be a relatively simple change in teaching technique in order to deal with the increasing complexities of war. The grasping, understanding, and mastering of adaptability will come through rigorous education and tough training early on—quality, not quantity—to produce adaptive leaders. Leaders' ability to be adaptable will guide decisions on how to accomplish their missions, while also helping them to recognize and compensate for differences in the temperament and ability of other Army officers, NCOs, and civilians through unit training and professional development. Adaptability will provide a stable support structure to infuse and sustain Army leaders' initiative in future operating environments.

Notes

I would like to thank the input and ideas of MAJ Kevin McNerney (USA ret.), CSM William Darwin (USA ret.), CSM Hugh Roberts (USA ret.), LTC Allen Gill (USA ret.), Mr. William S. Lind, and Dr. Bruce I. Gudmundsson.

1. Excerpts from the Army Campaign Plan, 2004, retrieved from campaignplan2004@us.army.mil.

2. http://www.almc.army.mil/ALOG/issues/NovDec00/News.htm.

3. Lisa Alley, "Basic Officer Leadership Course Gets Green Light" (March 2005), retrieved from http://www.aogusma.org/as/firstcall/Mar05/LeadershipCourse.htm.

4. This comes with discussions with 47 junior and middle grade officers from June 2005 to December 2007 who talk about more centralization of training management, as well as the pressure on field grades to do everything right when they are in their critical branch qualifying jobs as majors.

5. Input of Lieutenant Colonel John Sayen (USMC, ret.), January 7, 2007. John is an expert of force structure and organization, as well as equipment.

6. This is the definition the author came up with in September 2006.

7. The author is indebted to the work of members of the Asymmetric Warfare Group (AWG) and Kevin McEnery, who wrote an unpublished paper entitled "Changing Army Culture," which gave a great background to the influence of General DePuy on today's Army training and culture.

8. General William S. Wallace, "Victory Starts Here! Changing TRADOC to Meet the Needs of the Army," *Military Review* (Fort Leavenworth, KS: Combined Arms Center (CAC), May–June 2006).

9. Comes from discussions with 54 military and civilian personnel that work with or inside TRADOC.

10. Brigadier General David Fastabend and Mr. Robert Simpson, "Adapt or Die: The Imperative for a Culture of Innovation in the United States Army," *Army* (Arlington, VA: Association of the United States Army, February 2004); known herein as "Adapt or Die."

11. "Adapt or Die," p. 5.

12. Colonel Nicholas R. Marsella, "Effective Joint Training: Meeting the Challenges," *Parameters* (Carlisle, PA: U.S. Army War College, November–December 2004), downloaded from http://www.army.mil/professionalwriting/volumes/volume3/february_2005/1_05_3.html.

13. TRADOC Regulation 10-5, U.S. Army Training and Doctrine Command Organization and Functions, December 22, 2005.

14. TRADOC Regulation 350-70, "Training Development," March 9, 1999.

15. http://www.tradoc.army.mil/TPUBS/regs/r350-70/350_70_vi_2.htm.

16. Contains information on warrior or soldier ethos of the United States Army, retrieved from http://www.enlisted.info/about/warrior-ethos.shtml.

17. Walt Ulmer and Mike Malone, *Study of Professionalism* (Carlisle, PA: U.S. Army War College, 1970).

18. Colonel Richard M. Swain, "Selected Papers of William E. DePuy" (Fort Leavenworth, KS: Combat Studies Institute, January 1985).

19. Major Paul Herbert, "Deciding What Has to Be Done: General DePuy and the Creation of FM 100-5, Operations," *The Leavenworth Papers* (Fort Leavenworth, KS: Command and General Staff College, July 1988).

20. Colonel George Reed, "Systems Thinking and Senior Level Leadership," unpublished paper (Carlisle, PA: U.S. Army War College, August 2004).

21. Major Donald E. Vandergriff, "The Specter of Taylorism," *Army Times* (Springfield, VA: Army Times Publishing, November 2004).

22. Paul L. Savage and Richard Gabrial, "Turning Away from Managerialism: The Environment of Military Leadership," in *Military Review* (Fort Leavenworth, KS: U.S. Command and General Staff College, July 1980), p. 57.

23. David McCormick, *The Downsized Warrior: America's Army in Transition* (New York: New York University Press, 1998).

24. John C. Tillson, Merle L. Roberson, and Stanley A. Horowitz, *Alternative Approaches to Organizing, Training, and Assessing Army and Marine Corps Units* (Alexandria, VA: Institute for Defense Analysis, November 1992), p. 23: A band of excellence resembles a roller coaster of training ups and downs with the band representing the mean average of the surges of training that occur with constant inflow and outflow of personnel.

25. General Paul F. Gorman, *The Secret of Future Victories* (Fort Leavenworth, KS: Combat Studies Institute, June 18, 1979), pp. 56–57.

26. McCormick, pp. 120–156; Dave McCormick talks about the selections for early retirement boards used to help accomplish the Army's drawdown.

27. General Peter Schoomaker, "The Future of the United States Army." Remarks given at the American Enterprise Symposium, "The Future of the United States Army," April 11, 2005.

28. Both the Officer and NCO Army Training and Leader Development Panels' reports are at http://www.army.mil/features/ATLD/ATLD.htm.

29. Colonel Mike Galloucis, "Changing the Army Culture," unpublished memorandum, Army CSA Staff group (Washington, DC: Department of Army, February 13, 2005).

30. Quality assurance was extended into training to ensure that training regulations, policies, and doctrine were being followed, to include adherence to posted training schedules and mission training checklists.

31. Les Brownlee and General Peter Schoomaker, "Serving a Nation at War: Toward a Campaign Quality Army with Joint and Expeditionary Qualities," *Parameters* (Carlisle, PA: U.S. Army War College, 2004), p. 19.

32. Colonel Casey Haskins, "Commander's Mid-tour Assessment, 198th Infantry Brigade," unpublished memorandum (Fort Benning, GA: U.S. Army Infantry Center, August 13, 2007).

33. Haskins, "Commander's Mid-tour Assessment," pp. 1–2.

34. U.S. Army, *Field Manual 7-0 Training* (Fort Monroe, VA: U.S. Army TRADOC, June 2006), p. 3.

35. See Annex A for more details.

36. See Annex B and C for more details.

37. Asymmetric Warfare Group (AWG) located at Fort Meade, MD, was originally founded in 2005 as the IED Task Force, and came in to being as AWG in 2006. It takes the latest lessons learned in theaters of operations (war) and helps commanders and staffs translate these lessons into tactics, techniques, and procedures (TTPs). It has just recently begun teaching the CATC course, which has grown popular within a few short months with units and courses throughout the Army.

38. The author is indebted to the work of the Asymmetric Warfare Group, particularly by Command Sergeant Major Mike Cortes, Command Sergeant Major (ret.) Morgan Darwin, and Command Sergeant Major Hugh Roberts, as well as the positive command environment set by Colonel Robert Shaw, commander AWG, and Lieutenant Colonel Mike Richardson, commander Bake Squadron, AWG.

39. William M. Darwin, "Adaptability Learning Symposium," Asymmetric Warfare Group (Fort Meade, VA: AWG, December 11, 2007), slide 3.

40. Darwin, slide 7.

41. Based on numerous discussions with Command Sergeant Major Morgan Darwin (USA, ret.), November–December 2007.

42. Discussions with William Darwin, November 16, 2007.

43. Ibid.

44. Ibid.

45. Donald E. Vandergriff, "Review of Army Training," unpublished paper (Arlington, VA: Army TRADOC, September 2006), pp. 4–5.

46. Based on discussions with Bruce I. Gudmundsson regarding the German approach to leader development.

47. Discussions with William S. Lind, May 2008. Also based on lecture of Donald E. Vandergriff to the USMC Expeditionary Warfare School, September 2007.

48. Dr Robert Bjork, "How We Learn Versus How We Think We Learn: Implications for the Organization of Army Training," unpublished presentation, Science of Learning Workshop (Fort Monroe, VA: U.S. Army Training and Doctrine Command, August 1, 2006).

49. Training and Doctrine Command, *TRADOC Area of Interest 2: Leader Development and Professional Military Education* (Fort Monroe, VA: TRADOC, February 2006), p. 1.

50. Army Research Institute (ARI), Science of Learning Workshop (Fort Monroe, VA: TRADOC, August 2006).

51. U.S. Army Accessions Command, "Basic Officer Leaders Course (BOLC) Policy and Guidance," unpublished memorandum (Fort Monroe, VA: TRADOC, April 24, 2008), p. 4.

52. CPT Alec Barker, "Review of Raising the Bar," unpublished book review (Reston, VA: April 2008).

53. Donald E. Vandergriff, "Swift to Swiss: Tactical Decision Games and Their Place in Military Education," *Improved Performance Society Journal* (Rockville, MD: ISPI, February 2006).

54. Colonel Haskins did recognize the value of CATC after being introduced to it, and quickly used CATC as an asset to assist changes he was already implementing.

55. Listening to a lecture of Colonel Haskins to his leaders, September 17, 2007.

56. Haskins, "Commander's Mid-tour Assessment," p. 5.

57. Ibid., p. 6.

58. Ibid., p. 6.

59. "Adapt or Die."

60. Lieutenant General Walter Ulmer, "Creating and Assessing Productive Organizational Climates," Army War College Course Handout (Carlisle, PA: U.S. Army War College, 2002), p. 1.

CONCLUSION

Work Remains to Be Done

"It's the personnel system, stupid."

Secretary of the Army Thomas White, September 2002

The Army has identified some key weaknesses in its development of leaders, including weaknesses in cultural awareness, critical thinking, pattern analysis, adaptability, and self-awareness. Of critical note, observers noted a lack of understanding of irregular warfare, network skills, and how to teach. Also, it was deemed necessary to achieve a balance between the hard skills, math and sciences, versus so-called "soft skills," humanities, the area from which most of the leaders of the Army come. In sum, the Army is analyzing and proposing deep changes to almost every aspect of its soldier and leader professional development system. It is time for the institutional Army to catch up with the operational Army. Today, after six years of war, the experience of many leaders at the tactical level has evolved and improved.

According to Major Chris Coglianese, just returning from his second deployment to Iraq, Army units have improved significantly at the tactical level:

> There is practically no employment difference today in Iraq between light battalion and a heavy battalion, especially from the heavy battalion perspective. They have a couple dozen more guys, but a lot less in other ways (no engineers and limited organic maintenance to name just two things). All of our guys have a pretty substantial level of competency in "light infantry" skills. If you were to look at how our Soldiers in our tank companies (all companies were task organized to include the scouts attached to our main effort company) were outfitted, you could easily mistake them for an early 1990s "Delta" operator. One of our tank companies was our main effort and captured the most high-value individuals. We did night patrolling. We did ambushing as required.

> Bottom line—while we have atrophied many high-end tactical skills as an Army, our battalions (mine is very good, but not the only very good one to be sure) are pretty sophisticated at the counterinsurgency going on in Iraq (I cannot speak for Afghanistan).[1]

The fundamental problem remains untouched, however. While many vast improvements are beginning to be made to how soldiers and leaders are recruited and then trained and developed, the Army continues to deal with short-term adjustments to an already over-taxed, legacy personnel management system designed to produce soldiers and leaders trained to fight in linear war. Proposed reforms to Army culture still avoid changing the system's legacies, which also serve as the four pillars holding up the cultural structure.

The four legacy pillars of today's culture are

1. The up-or-out promotion system (instituted with the 1916 Naval Personnel Act, and further institutionalized by the Officer Personnel Act [OPA] in 1947); finally accepted as the only way to manage officers in the Defense Officer Personnel Act of 1980 or DOPMA;
2. Quantity-based vs. quality-based officer accessions system (institutionalized by the Morrill Act of 1862 and the National Security Act of 1916);
3. Centralized control of individual performance appraisal and selection systems;
4. A top-heavy officer corps and too many headquarters.

As long as these legacies of today's Army culture remain invulnerable, the service will evolve only slowly, or not at all, and therefore will have trouble in recruiting, developing, and retaining adaptive leaders and soldiers. These pillars of current Army culture focus on short-term gratification that surreptitiously contradicts the Army value of selfless service. This forces anyone serious about understanding what it takes to move the Army from a linear system to one that deals with complexity to question whether the service is truly serious about its professed values.[2]

Such a system places all soldiers—but particularly officers—in a climate of conflict between Army rhetoric and reality. They are confronted with the hard choice almost every day: to do what is right, and seek the attainment of hard to earn and, in many cases, long-term, professionalism, while possibly suffering negative consequences; or to do what the career system demands, by climbing short term career ladders without slipping up. In the end, the latter choice will ensure career success, but at a price of being suborned by the system itself.

In a significant study sponsored by the Army War College in 2003, Colonel Stephen Jones explained how the culture can fail to align with proposed advances in leader development:

> Several adverse trends in command climate have persisted in the Army for nearly 30 years, perhaps because, in practice, the officer culture emphasizes short-term mission accomplishment more than long-term organizational

growth, or because Army systems reinforce individual performance rather than organizational effectiveness. Either emphasis, if true, detracts from combat readiness.

Compounding the problem, Army leaders are not taught how to assess or improve command climate, nor are they rewarded when they do so. Army organizations, officers, and soldiers deserve better. Cultural norms and counterproductive evaluation, leader development, and accountability systems are at the root of the U.S. Army's problems regarding organizational (command) climate. Absent a shift in cultural emphasis and adjustment of systems to reinforce the change, command climate will continue to suffer; and unit effectiveness, morale and trust, retention, and commitment will continue to be significantly degraded as the war continues.[3]

Soldiers and families are tremendously proud of their service, and they enjoy the camaraderie of a military life. They are enthusiastic about the positive impact they're making on mission success and on those they lead. Several issues are impacting and will impact recruiting and retention today and in the future, and are only being boxed in by the legacies mentioned earlier.

The mandated 15-month deployments are extremely taxing on soldiers and families. Time away from families (*and lack of time to start a family*) is a major reason for soldiers leaving the Army. Unit leaders cite a decrease in operational effectiveness after 10 to 12 months deployed, and a significant decrease in effectiveness after a year deployed. Casualties sustained after one year (deployed) are also significantly more emotional for the unit, soldiers, and families.

Lengthening tours or changing expectations after deployment kills morale. Families plan events, make financial commitments, and emotionally tie themselves to a set time apart. Families measure deployment in terms of "life events" (birthdays, anniversaries, holidays) and missing more than one of these consecutively is demoralizing. Soldiers and families want more than 12 months' dwell. Soldiers believe training is too robust *too soon* after returning from deployment. Field training, training off post, and professional development demands for noncommissioned officers (NCOs) and officers (OES/Non Commissioned Officer Education System [NCOES]) also cause increased stress on families during dwell periods.

Junior officers and NCOs want to continue their professional development despite "stop move" instructions. Captains want to attend their Career Course "on time" (with contemporaries), NCOs want to attend NCOES schools, and both groups feel "left behind" when they don't. The top-heavy and over-strength assignment of captains in deploying/deployed brigade combat teams (BCTs) adds to junior officer frustration and causes many to feel underutilized with regard with their education and skill sets. Finally, junior officers and unit leaders both want more say in future assignments, and feel they have little influence due to the demands on Human Resource Command (HRC).

Assignment to *external* Mission Training Teams (MiTTs—advisory teams) is a major cause of dissatisfaction among junior officers and is perceived as a personally and professionally unrewarding assignment. Officers perceive MiTTs as ad-hoc, and do not believe MiTT count as much as an assignment to a unit.

Another problem, already touched upon in Chapter 3, "U.S. Society's Impact," is that soldiers perceive the Army as bearing an inordinate burden for the war without commensurate recognition or compensation. Many soldiers have a perception that other services do a better job of rewarding their servicemen and women because of their shorter deployment periods and higher reenlistment (and deployment) bonuses. Soldiers are also frustrated that there is not more effort or participation by other governmental agencies throughout Iraq and Afghanistan.

All junior officers do not feel adequately compensated or recognized for their hard work. High promotion rates and Critical Skills Retention Bonus (CSRB) incentives fell short of expectations and changed few minds regarding their decision to stay.[4] All junior officers are not being adequately counseled or mentored by their superiors and many do not aspire to the lifestyle they see in their senior leaders. Junior officers, particularly captains, get easily frustrated by any lack of caring among field grade leaders—majors through colonels—and by not knowing where they stand in their career or what opportunities lay in their future. Leadership is clearly a cause in many retention decisions.

Fifteen-month combat tours have caused increased physical and emotional strain on soldiers and their families. While the unit cyclic program the Army has mandated is a good idea, the Army must have more units to make it work. This strain is further compounded in those with previous combat deployments, and those facing subsequent deployments shortly after their return from theater. As a result, many soldiers, NCOs, and officers are "on the fence" regarding continuing service in the Army. Despite professional fulfillment, a personal *commitment to family* and/or frustration with a lack of change in a system that needs it, overrides decisions for many to stay in the Army.

The Army Promotion System

The Army pillars include, first and foremost, the "up-or-out" promotion system. Up-or-out, the requirement to achieve periodic promotion or leave military service, is the strongest part of the foundation and, in turn, supports the other three pillars. Knock down the up-or-out policy and the other pillars will follow. Up-or-out is a unique U.S. alternative to European military organizations' professional entrance requirements, a step that satisfied anti-militarism sentiments in American civilian society.

Up-or-out was also an acceptable replacement in the face of Congress' opposition to the rigorous entrance requirements that the Prussians created to restrict and control access to their officer corps. Creating tough professional entrance requirements based solely on aptitude for military leadership in America was almost impossible in Emory Upton's day in the late 1800s. The glow of victory after World War II, with the United States having defeated the warlike cultures of

Germany and Japan, also made it appear unlikely and unnecessary to reform the system. However, there are aspects of the Germans' education of junior leaders in cognitive and character development that merit examination.

Transformation has normally been quicker and more complete for defeated armies unburdened by the legacy of a recent victory. (Nevertheless, the goal of this monograph is not to duplicate a specific doctrine or historical period with either a victorious or defeated Army.) The German approach established the concepts of both professional and individual trust early in the training process, as every officer—through a shared bonding experience—had to negotiate and pass through the same tough standards. This accomplished the creation of a genuine bond of trust within a small but professional officer corps, and it freed the officers to focus on mastering their *professions* instead of their *careers*.[5]

The U.S. alternative, while lowering the rigor of initial entrance requirements into the officer corps, created a continual career-long endurance test for those who joined it. This accompanied the Army's move at the beginning of the twentieth century to the philosophy of "Taylorism," which focused on "producing" large numbers of officers in the event of a major war.

The U.S. officer accessions process became a rough, never-ending road of short-term performance measures that are far from the standards required for winning on the 4GW battlefield. Up-or-out was created to prevent the appearance of elitism (which other civilian professions such as medicine, engineering, and law subtly claim), while proclaiming the façade of equity for all who possess the will to climb.

In fact, up-or-out creates something much worse, as personnel expert and author LTC Harry Bondy proclaims through his extensive study of personnel management systems:

> [The up-or-out policy] does little to improve team performance because individuals do not have a measurable effect on productivity. Legitimacy and commitment suffer because almost everyone not promoted to senior officer and non-commissioned member rank . . . is dissatisfied with the current system. Those who are promoted and in control dismiss the dissatisfaction as "sour grapes."[6]

LTC Ike Wilson, PhD, who is currently assigned as a professor at West Point, points out an additional result of the up-or-out promotion system in a U.S. Army School of Advanced Military Studies manuscript in 2003:

> This author's criticism begins where [Maj. Donald E.] Vandergriff's ends: the negative and unintended effects OPA 47 and the present officer management system is levying on officer experienced-based learning. The legacy of OPA 47 continues to enforce the "competitive ethic" that derives from a centralized promotion system that, by design and implementation, "defines success" upon a very small set of critical career-enhancing positions and experiences—command, aide-de-camp (to a general officer), and "key and essential" staff

> officer positions, such as Battalion and Brigade operations officers (S-3s) and executive officers (XOs). . . . The argument that a "one size must fit all" approach to officer promotions, assignments, and experiencing is wrought with problems.[7]

The second pillar is an officer "production" system that uses never ending yearly cohorts (the grouping of all officers, of various skills, needed in the Army for that fiscal year), which, in turn, creates a "promotion tournament" that undermines trust between officers and encourages negative competition—as if a new automobile model must be turned out yearly in the competitive automotive industry.

The third pillar is the centralized control of promotions and selections based upon individual evaluation systems. In practice, these create the psychological conflicts that counter the values the Army espouses. Bondy continues:

> Modern management techniques do not always build good social capital and discipline. Research has shown that individual performance appraisal and selection systems, for example, are inaccurate, unscientific, and prone to subgroup subversion. More than half of any rating variance is due to "idiosyncratic rater effects" such as how much the rater likes the ratee; whether they have similar personalities; their views on performance; stereotypes on gender, race and ethnicity; self-interest; sub-group factional interests; and variations in work context, which are significant in the military. Most importantly, centralized transfer and promotion queues lead to frequent, expensive postings that reduce social capital, erode trust, and add to careerist credentialism. The annual promotion "tournament" shifts people between units, as if robbing Peter to pay Paul, primarily to reward the winners. This does little to improve team performance because individuals do not have a measurable effect on productivity.[8]

The final pillar is the top-heavy officer corps. It, too, has become an accepted truth: The only way of doing business is for the Army and military to possess a high ratio of officers, particularly middle or field grades, to junior grades and soldiers. It evolved to the point that both headquarters and laws have been created to justify the officer bloat. However, a top-heavy officer corps is counter to an effective Army trying to operate in a 4GW environment, as veteran defense analyst Franklin C. Spinney points out:

> The increasing proportion of officers coupled with the downward trend in force size since the early 1950s means that command opportunities have declined over the long term, and command experience has decreased correspondingly, because command tours have become shorter as we shoveled people through platoon- and company-level commands to feed the voracious appetite for officers on the proliferating variety of staffs. This evolution naturally pushed decisions up the chain of command to higher ranks and led to increasing centralization as officers on staffs found it increasingly difficult to

> justify their existence. Couple that with mobilization-centric personnel policies like the up-or-out promotion system and you have the ingredients for a risk-averse, ticket-punching careerist culture, where self-interest takes precedence over organizational purpose. Not surprisingly, younger people, particularly captains, with their futures still before them, are saying, "This is BS, I'm outta here!"[9]

What the Army must address, and fix, are theses untouchable legacies that lie hidden underneath many cosmetic proposals to educate and train leaders. The problem is that the legacies, particularly the top-heavy officer corps and the up-or-out promotion system, are valued by those within the Army who have the chance to lose the most as they rapidly climb to the top as "unmentionable untouchables." The Army's strategic leaders must have a long-range vision to challenge the legacy systems in order to replace them with pillars that support a culture that encourages and protects adaptive leaders.

Legacies Impact Organizational Effectiveness

If anything, recent successes have allowed the Army to continue unknowingly toward a more bureaucratized and centralized organization in light of all the proclaimed and real attempts at implementing adaptability. The four pillars of the cultural foundation are the most potent and subtle social control mechanisms in the Army's previously discussed legacies. These legacies, defined by laws and policies, as well as the culture's criteria of success, "have the greatest impact on demonstrating and teaching the values of the organization."

Retention of these legacies provides the primary power levers for changing or maintaining culture. These legacies, particularly promotions and selections tied to the up-or-out promotion policy and individual evaluation systems, presented as inherently fair, determine awards and access to positions of influence and control. They provide specific instructions when tasking subordinates due to an obsession with certainty. The individual as well as the "system" carefully monitor the execution of their instructions, and track all activities and outcomes with the finest attention to detail. Unfortunately, "professional systems and structures are not very adaptable."[10]

If the Army is to become adaptive, and a "learning organization," it must ensure that its personnel system supports its move to adaptability—in people and institutions, and not the other way around, where a retained personnel system limits and diminishes any implementation that results in adaptability. The thread of evolutionary adaptability must exist everywhere. It starts with doctrine and strategic leaders and filters down to daily activities, threads through policies and beliefs, winding its way from the institutional Army to those forces deployed in the conduct of an array of possible future missions. An environment must be in place to support and nurture the adaptability the Army says it wants in its leaders and soldiers.

Legacies combine to produce an untrusting environment between superiors and subordinates. This, in turn, necessitates increased supervision, setting the stage for micromanagement. Distrust of subordinates hinders delegation of authority, thereby increasing the amount of information that must be processed at all levels, which, in turn, slows down the Observe Orient Decide Act (OODA) loop (or Boyd decision cycle). Meanwhile, the increased demand for decision-making at the lower levels in a 4GW operating environment force strategic commanders to legislate explicit directions (rules and regulations). As the volume of regulations heaped upon subordinates increase, the overall organization becomes a more rigid mechanical and bureaucratic edifice.

"Method driven orders" (evolving from increased regulation) increase the need for explicit communication between superiors and subordinates. A greater overlap of responsibility occurs, leading to confusion, friction, and competition for authority between superiors and subordinates. Mistrust ensures that tactics are mandated from one level down to the next.

This situation leads to underdeveloped tactical expertise. Commanders treat subordinates less like "thinking beings" and more like slavish, robotic servants. The creative and intellectual ability of individual commanders is underutilized, generating dissatisfaction, cynicism, frustration, and low morale. Tactical commanders, given narrow authority, become hesitant to make decisions. The continuous need to obtain permission to execute plans stifles initiative. As they can no longer fluidly adapt to changing circumstances, subordinates become trapped in rigidly defined roles like cogs in a machine and respond ineffectively to a rapidly changing enemy.

Short-Term Fixes Now

While beginning to address the four legacies mentioned, which will take time, some short-term fixes can be executed by the Army chain of command that will assist in recruiting and retention by stating, "We care and are listening." As Kotter states, these would be short-term victories by those selling change to the entire organization.

The entire organizational leadership must continue to highlight the prestige, importance, and gratitude for military service, using it as a selling point in both recruiting and retention. It should encourage partnering with local communities and national leaders in this effort, as mentioned previously. The new president of the United States must take the lead in this effort beyond mere rhetoric, while members must do more than just put yellow ribbons on their SUVs.

As soon as feasible, deployment policy needs to be changed back to 12 months. Many soldiers/leaders would re-enlist/accept retention payments mentioned in Chapter 5 if deployments were shortened, and leaders must remain loyal to announced tour durations as much as possible. If deployment timelines change, they must announce changes through the chain of command as soon as possible, and explain why as well. As soon as possible, the Army should lengthen "dwell" time in the United States or unit home bases to 15 months. Once the decision is made to do this, advertise commitment to dwell time of 2:1 (short term, two dwell times to every one deployment) and 3:1 (long term, three dwell periods to one deployment). Senior leaders must ruthlessly protect family time on redeployment, and think creatively of ways to train/prepare for future deployments while protecting soldiers' time with their families.

On top of addressing time home, also mandate timely attendance at Captains Career Course for leaders, and provide officers and NCOs continued opportunities to develop as individuals, even while deployed. Flattening the organization will limit "over-strength" assignment of captains in deploying BCTs, and closely coordinate inbound assignments from the Army's HRC to prevent overpopulation. If officer assignments create overpopulation of captains in deployed units, return officers from theater to attend their Career Courses and other professional development schools. Allow junior officers more input in their future assignments. Reinstitute a web-based menu of assignment opportunities, and allow more flexibility for officers to swap assignments based on their desires. Finally, study alternatives to the current external MiTT assignment process to one where MiTTs are assigned internal to deploying BCTs.

Continue to explore creative ways to compensate soldiers participating in long and subsequent deployments: Increase hazardous duty and family separation pay, explore nonchargeable and expanded mid-tour and block leave programs, continue robust support for families while spouses are deployed, and continue to improve housing, medical care, and dental benefits for families. Seek Department of Defense review of deployment benefits to ensure parity across all services, and publicize results.

Finally, the Army must improve chain-of-command counseling, mentoring, and leading junior officers and NCOs. Provide junior officers with feedback on their performance, and talk to them about opportunities in their future. Continue to put our best officers in command, and recognize our very best with second command, teaching assignments, early promotions, incentive pay, choice of duty position or post, and increased educational opportunities. Institute a 360-degree "rater/senior rater feedback" system whereby senior officers (battalion and brigade commanders) are evaluated on their leadership/mentoring efforts for their subordinates.

Change the Pillars and Culture Will Evolve

> "We can agree that what will be needed are more flexible and adaptable officers and a more flexible and adaptable officer corps."
>
> "Secretary of Defense 2003 Summer Study: The Military Officer in 2030"
>
> Defense Department, Office of Net Assessment

Leaders adapt to the conditions of war under the "right" command climates. The conditions of war demand good leaders and commanders to put aside and ignore the pillars of promotion (described earlier) that drive the internal organizational culture.

What happens when the Army does not adapt to 4GW? As the Army tries to change everything *but* its culture, a strategic mismatch occurs with an enemy, who continually evolves to its changing operating environment. Boyd said that effective armies throughout history applied a "people first, then ideas, and finally hardware" approach to adapting to environmental changes.

In previous generations of war, the United States relied on a "technology over people" approach that might have led to a tactical defeat or a setback initially, but the nature of warfare and the global environment of the time allowed the nation time to overcome the mistake. For example, in World War II, once the arsenal of democracy got its production lines going, it overwhelmed the enemy with materials and firepower. Another example is Vietnam, where material might and firepower failed to win the war, but the United States was able to recover from the defeat because the lack of globalization and the technology of the day kept the defeat isolated to that part of the world.

Today, setbacks on the 4GW battlefield have broader and deeper strategic implications such as the strengthening of the Iraqi insurgency because of the prisoner abuse at the Abu Ghraib prison. Lack of leadership at Abu Ghraib and the policy of treating all Arabs as suspected "terrorists" served to recruit thousands within and outside Iraq to join the insurgency. A culture that understands the evolution of conventional warfare into 4GW approaches warfare differently. The nation may not be able to recover from a tactical setback today.

In order to set the conditions for changing the Army's culture as well as those for its new operating environment, it is critical to recognize how change is implemented and where the biggest impact falls. Trust of those at the top of the organization—the strategic leaders—erodes as rhetoric continues, yet nothing really changes. When this happens, trust is replaced by frustration, sarcasm, and finally a lack of commitment and acts of self-preservation instead of selfless service. Jonathan Shay observes and chronicles similar cultural occurrences during the Vietnam War in his book, *Achilles in Vietnam* (Scribner, 1994).

Part of the cultural problem involves hierarchical values. It is questionable today whether those at the top of the hierarchy really value something other than those values that led to their own advancement to the higher ranks. Chris Argyris, a leadership and cultural expert at Harvard University, calls the dichotomy of "rhetoric contrasting reality" "espoused theory," which often differs from the "theory in use."

Pinpointing and understanding the difference between these "theories of action" is a key to beginning a plan to evolve the culture. A lack of this understanding is what frustrates members of the organization who know change is necessary and want to implement it, but over time do not see it occur. They are normally not in a position to impact or influence the change and feel they do not have a voice in the process.

Army culture has been learned both in psychological and social contexts. Hence, culture needs to be unlearned, and then relearned. Many popular management books claim that "re-engineering" culture is difficult because (1) it is not an engineering problem and (2) no one knows what the outcome of changing the culture should be. The members of the organization should be suspicious of those inside the culture (especially "those in power") who define the "right" changed culture. After all, they themselves are products of the current culture, raising the question of whether they can be objective in changing it. But the existing operating environments seen springing up all over the world (Afghanistan and Iraq) must be addressed with a new leader paradigm.

The emergence of 4GW demands rapid adaptive decisions from all levels, while understanding the commander's intent, not two, but three levels higher—and even further, up to the strategic level. 4GW also means that knowing how to fight is not enough.

Adapting the culture to the demands of 4GW calls for a bottom-up evolution. The ultimate goal is for those at the top to have the strategic vision to "exploit," highlight, and award innovations that will move cultural evolution even further. Those at the bottom who must adapt have an idea of what the Army culture should be and are seeing some signs that changes are occurring in reaction to 4GW environments. This happens because junior leaders—officers and NCOs—begin to understand that the key to fighting and winning in a 4GW environment is to understand the three aspects of war—the physical, the mental, and the moral, with the moral being the most important. This stands in sharp contrast to an "American way of war" that emphasizes the physical aspect of war (2GW), and the destruction and counting (measuring) of targets.

As mentioned at the beginning of this book, many people in the Army are beginning to understand the dominance of the moral side of war, and this started at the top with former Army Chief of Staff General Peter Schoomaker, as well as other Army officers. Those who understand the moral side of war use innovations that include nonlethal means to solve tactical problems. This ability to adapt easily supports the type of strategy the United States will have to use to be successful in 4GW. At the very least, these leaders do not negatively impact ongoing missions.

The Army is just now addressing the mental aspect of war surrounding today's culture from top to bottom. An example of a top-down effort is Schoomaker's insistence that officers and NCOs read books such as Lieutenant Colonel John Nagl's *Learning to Eat Soup with a Knife: Counterinsurgency Lessons from Malaysia and Vietnam*. A bottom-up example is the circulation of "lessons learned" by captains who have served in Iraq using the Internet site www.companycommand.com. To produce adaptive leaders, Army institutions have to become adaptive themselves to evolve current lessons and techniques into curricula as quickly as possible. Today's leader-development curricula are out of date by the time of publication.

Notes

1. E-mail from Major Chris Coglianese to author, January 18, 2008.

2. Harry J. Bondy, "Postmodernism and the Source of Military Strength in the Anglo West," *Armed Forces and Society*, vol. 31, issue 1 (College Park, MD: November 2004).

3. Colonel Stephen Jones, "Improving Accountability for Effective Command Climate: A Strategic Imperative," *Strategic Studies Institute* (Carlisle, PA: U.S. Army War College, September 2003), p. v.

4. Army News Service, "20k Bonus, Other Incentives Available for Some Active-Duty Captains," retrieved from http://www.globalsecurity.org/military/library/news/2007/05/mil-070517-arnews01.htm.

5. For the best two descriptions of the Prussian/German officer accessions process, see Bruce I. Gudmundsson, *Stormtroop Tactics: German Innovation 1914–1918* (New York: Praeger, 1995); James Corum, *Roots of Blitzkrieg* (Lawrence, KS: University Press of Kansas, 2000). For a good summary, see http://www.d-n-i.net/vandergriff/rha/sld034.htm and slide 35.

6. LTC Harry J. Bondy, "New Regiments, New Specialist Corps, and a New General Staff," *Journal of Military and Strategic Studies* (Canadian Defense and Foreign Affairs Institute, Winter 2004), pp. 3–5.

7. Major Isaiah Wilson III, "Educating the Post-modern U.S. Army Strategic Planner: Improving the Organizational Construct" (Fort Leavenworth, KS: U.S. Army School of Advanced Military Studies, February 2003), pp. 43–44.

8. Author e-mail correspondence with LTC Harry J. Bondy (Canadian Army), April 2005. Also see Bondy, "New Regiments, New Specialist Corps, and a New General Staff," pp. 3–5; available at http://www.jmss.org/2004/winter/articlesbody5.htm. Bondy and the author have shared many facts and ideas over the past few years. Many of the factors that affect the U.S. Army also affect the Canadian Army.

9. Franklin C. Spinney and LTC John Sayen, "Sayen Report: Officer Bloat Creates a Shortage of Captains," *Defense and National Interest,* July 16, 2000; retrieved from http://www.d-n-i.net/fcs/comments/ c372.htm.

10. William Bell, "The Impact of Policies on Organizational Values and Culture," paper presented at the Joint Services Conference on Professional Ethics, January 1999.

Index

About the Author

DONALD VANDERGRIFF (Major, U.S. Army retired) is currently a consultant on leader development, adaptability, and how to create and sustain learning organizations to the U.S. Army. The U.S. Marines and Navy SEALs, as well as several corporate organizations, have sought his insights on adaptability. He is the author of 50 articles, as well as three books.

Recent Titles in
Contemporary Military, Strategic, and Security Issues

Military Base Closure: A Reference Handbook
David S. Sorenson

Military Education: A Reference Handbook
Cynthia A. Watson

Strategic Defense in the Nuclear Age: A Reference Handbook
Sanford Lakoff

Intelligence and National Security: A Reference Handbook
J. Ransom Clark

Military Reform: A Reference Handbook
Winslow T. Wheeler and Lawrence J. Korb

The U.S. Military's "Don't Ask, Don't Tell" Policy: A Reference Handbook
Melissa Sheridan Embser-Herbert

Prisoners of War: A Reference Handbook
Arnold Krammer

Nation-Building and Stability Operations: A Reference Handbook
Cynthia A. Watson

Military Transformation and Modern Warfare: A Reference Handbook
Elinor Sloan

Information Operations—Doctrine and Practice: A Reference Handbook
Christopher Paul

The National Guard and Reserve: A Reference Handbook
Michael D. Doubler

Returning Wars' Wounded, Injured, and Ill: A Reference Handbook
Nathan D. Ainspan and Walter E. Penk, editors